Data Wrangling with R

Load, explore, transform and visualize data for modeling with tidyverse libraries

Gustavo R Santos

BIRMINGHAM—MUMBAI

Data Wrangling with R

Group Product Manager: Reshma Raman

Publishing Product Manager: Apeksha Shetty

Senior Editor: Sushma Reddy

Technical Editor: Rahul Limbachiya

Copy Editor: Safis Editing

Project Coordinator: Farheen Fathima

Proofreader: Safis Editing

Indexer: Tejal Daruwale Soni

Production Designer: Nilesh Mohite

Marketing Coordinator: Nivedita Singh

First published: January 2023

Production reference: 1310123

Published by Packt Publishing Ltd.

Livery Place

35 Livery Street

Birmingham

B3 2PB, UK.

ISBN 978-1-80323-540-0

`www.packtpub.com`

To my wife, Roxane, my other half.

To my children, Maria Fernanda and Marina,

who remind me every day how much I still need to learn.

- Gustavo R Santos

Contributors

About the author

Gustavo R Santos has worked in the technology industry for 13 years, improving processes, analyzing datasets, and creating dashboards. Since 2020, he has been working as a data scientist in the retail industry, wrangling, analyzing, visualizing, and modeling data with the most modern tools such as R, Python, and Databricks. Gustavo also gives lectures from time to time at an online school about data science concepts. He has a background in marketing, is certified as a data scientist by the Data Science Academy, Brazil, and pursues his specialist MBA in data science at the University of São Paulo.

About the reviewers

Chirag Subramanian is an experienced senior data scientist with more than six years of full-time work experience in data science and analytics applied to catastrophe modeling, insurance, travel and tourism, and healthcare industries. Chirag currently works as a senior data scientist within the global insights department at Walgreens Boots Alliance, a Fortune 16 company. Chirag holds a master of science degree in operations research from Northeastern University, Boston. He is currently pursuing his second master of science degree in computational data analytics from Georgia Tech, a top-10 globally ranked school in statistics and operational research by QS World University Rankings. His hobbies include watching cricket, playing table tennis, and writing poetry.

Love Tyagi is the director of data science at one of the top biotech companies in the USA. He completed his bachelor's in engineering from Delhi, India, and pursued his master's in data science from George Washington University, USA. Love enjoys building solutions around data analytics, machine learning, and performing statistical analysis. He is, at the core, a data person, and wants to keep improving the way people learn about data, algorithms, and the maths behind it. And that's why, whenever he has time, he likes to review books, papers, and blogs.

When not working, he enjoys playing cricket and soccer and researching new trends in cooking, as he is also a part-time chef at home. He also loves teaching and someday wants to go back to teaching and start his own institution for learning AI in the most practical and basic way possible.

Table of Contents

3

Basic Data Visualization   51

Part 2: Data Wrangling

4

Working with Strings   79

5

Working with Numbers   111

6

Working with Date and Time Objects   131

7

Transformations with Base R   155

8

Transformations with Tidyverse Libraries   177

9

Exploratory Data Analysis   215

Part 3: Data Visualization

10

Introduction to ggplot2   245

11

Enhanced Visualizations with ggplot2   269

12

Other Data Visualization Options   291

Part 4: Modeling

13

Building a Model with R   305

14

Build an Application with Shiny in R   333

Conclusion   349

Index   353

Other Books You May Enjoy   360

Preface

Data Science is a vast field of study. There is so much to learn about and, every day, more and more is added to this pile. It is fascinating, for sure, the way we can analyze data and extract insights that will serve as a base for better decisions. The big companies have learned that data is what can take them to the next level of business achievement and are leading the way by building strong data science teams.

However, just data by itself is not the answer. It is like crude oil: out of it, we can make plenty of things, but just that black liquid from the ground won't serve us very well. So, raw data is something, but when we clean, transform, and analyze it, we are transforming data into information, and that brings us the power to make better decisions.

In this book, we will go over many aspects of data wrangling, where we will learn how to transform data into knowledge for our business. Our chosen programming language is R, an amazing piece of software that was initially created as a statistical program but became much more than that. If we know what we need to achieve, getting there is just a matter of finding the right tools. Many of those tools are in this book.

Who this book is for

This book is written for professionals from academia and industry looking to acquire or enhance their capabilities of wrangling data using the R language with RStudio. You are expected to be familiar with basic programming concepts in R, such as types of variables and how to create them, loops, and functions. This book provides a complete flow for data wrangling, starting with loading the dataset in the IDE until it is ready for visualization and modeling. A background in **science**, **technology**, **engineering**, and **math** (**STEM**) areas or in the data science field are not required but will help you to internalize knowledge and get the most from the content.

What this book covers

Chapter 1, *Fundamentals of Data Wrangling*, will introduce this book's main theme, explaining what data wrangling is and why and when to use it. In addition, it also shows the main steps of a data science project and covers three well-known frameworks for data science projects.

Chapter 2, *Load and Explore Datasets*, provides different ways to load datasets to RStudio. Every project begins with data, so it is important to know how to load it into your session. It also begins exploring that data to familiarize you with exploratory data analysis.

Chapter 3, Basic Data Visualization, is the first touch point with data visualization, which is an important component of any data science project. In this chapter, we will learn about the first steps to creating compelling and meaningful graphics using only the built-in library from R.

Chapter 4, Working with Strings, starts our journey of learning about the wrangling functions for each major variable type. In this chapter, we study many possible transformations with text, from detecting words in a phrase or in a dataset to some highly customized functions that involve regular expressions and text mining concepts.

Chapter 5, Working with Numbers, comprises the transformations for numerical variables. The chapter covers operations with vectors, matrices, and data frames and also covers the apply functions and how to make a good read of the descriptive statistics of a dataset.

Chapter 6, Working with Date and Time Objects, is where we will learn more about this fascinating object type, date and time. It introduces concepts from the basics of creating a date and time object to a practical project that shows how it can be used in an analysis.

Chapter 7, Transformations with Base R, is the core of the book, exploring the most important transformations to be performed in a dataset. This chapter covers tasks such as slicing, grouping, replacing, arranging, binding data, and more. The most used transformations are covered here and mostly use the built-in functions without the need to load extra libraries.

Chapter 8, Transformations with tidyverse Libraries, follows the same idea as *Chapter 7*, but this time, the transformations are performed with **tidyverse**, which is a highly used R package for data science.

Chapter 9, Exploratory Data Analysis, is all about practice. After going over many transformation functions for different types of variables, it's time to put the acquired knowledge into practice and work on a complete exploratory data analysis project.

Chapter 10, Introduction to ggplot2, introduces the visualization library, **ggplot2**, which is the most used library for data visualization in the R language, given its flexibility and robustness. In this chapter, we will learn more about the grammar of graphics and how ggplot2 is created based on this concept. We will also cover many kinds of plots and how to create them.

Chapter 11, Enhanced Visualizations with ggplot2, covers more advanced types of graphics that can be created with ggplot2, such as facet grids, maps, and 3D graphics.

Chapter 12, Other Data Visualization Options, is where we will see yet more options to visualize data, such as creating a basic plot in **Microsoft Power BI** but using the R language. We will also cover how to create word clouds and when that kind of visualization can be useful.

Chapter 13, Build a Model with R, is all about an end-to-end data science project. We will get a dataset and start exploring it, then we will clean the data and create some visualizations that help us to explain the steps taken, and that will lead us to the best model to be created.

Chapter 14, Build an Application with Shiny in R, is the final chapter, where we will take the model created in *Chapter 13* and put it in production using a web application created with Shiny for R.

To get the most out of this book

The get the most out of the content presented in this book, it is expected that you have a minimum knowledge of object-oriented programming (creating variables, loops, and functions) and have already worked with R. A basic knowledge of data science concepts is also welcome and can help you understand the tutorials and projects.

All the software and code are created using RStudio for Windows 10, and if you want to code along with the examples, you will need to install R and RStudio on your local machine. To do that, you should go to `https://cran.r-project.org/`, click on **Download R for Windows** (or for your operating system), then click on **base**, and finally, click on **Download R-X.X.X for Windows**. This will download the R language executable file to your machine. Then, you can double-click on the file to install, accepting the default selections.

Next, you need to install RStudio, renamed to Posit in 2022. The URL to download the software is found here: `https://posit.co/download/rstudio-desktop/`. Click on **Download** and look for the version of your operating system. The software has a free of charge version and you can install it, accepting the default options once again.

The main libraries used in the tutorials from this book are indicated as follows:

Software/Library	Version
R	4.1.0
RStudio	2022.02.3+492 for Windows
Tidyverse	1.3.1
Tidytext	0.3.2
Gutenbergr	0.2.1
Patchwork	1.1.1
wordcloud2	0.2.1
ROCR	1.0-11
Shinythemes	1.2.0
Plotly	4.10.0
Caret	6.0-90
Shiny	1.7.1
Skimr	2.1.4
Lubridate	1.8.0
randomForest	4.7-1
data.table	1.14.2

To install any library in RStudio, just use the following code snippet:

```
# Installing libraries to RStudio
install.packages("package_name")
# Loading a library to a session
library(package_name)
```

In R, it can be useful to remind yourself of, or have in mind, these two code snippets. The first one is how to write for loops. We can write it as, for a given condition, execute a piece of code until the condition is not met anymore:

```
for (num in 1:5) {
    print(num)
}
```

The other one is the skeleton of a function written in R language, where we provide variables and the code of what should be done with those variables, returning the resulting calculation:

```
custom_sum_function <- function(var1, var2) {
    # Function code
    my_sum = sum(var1 + var2)
    return(my_sum)
}
```

If you are using a digital version of this book, we advise you to type the code yourself or access the code from the book's GitHub repository, preventing any potential errors with code broken due to copy and paste.

Download the example code files

You can download the example code files for the tutorials contained in this book from GitHub at https://github.com/PacktPublishing/Data-Wrangling-with-R. If there are any changes to the code, it will be updated in the GitHub repository.

Conventions used

There are a number of text conventions used throughout this book.

Code in text: Indicates code words in text, database table names, folder names, filenames, file extensions, pathnames, dummy URLs, user input, and Twitter handles. Here is an example: "Mount the downloaded WebStorm-10*.dmg disk image file as another disk in your system."

A block of code is set as follows:

```
html, body, #map {
  height: 100%;
  margin: 0;
  padding: 0
}
```

When we wish to draw your attention to a particular part of a code block, the relevant lines or items are set in bold:

```
[default]
exten => s,1,Dial(Zap/1|30)
exten => s,2,Voicemail(u100)
exten => s,102,Voicemail(b100)
exten => i,1,Voicemail(s0)
```

Any command-line input or output is written as follows:

```
$ mkdir css
$ cd css
```

Bold: Indicates a new term, an important word, or words that you see onscreen. For instance, words in menus or dialog boxes appear in **bold**. Here is an example: "Select **System info** from the **Administration** panel."

> **Tips or important notes**
> Appear like this.

Get in touch

Feedback from our readers is always welcome.

General feedback: If you have questions about any aspect of this book, email us at customercare@ packtpub.com and mention the book title in the subject of your message.

Errata: Although we have taken every care to ensure the accuracy of our content, mistakes do happen. If you have found a mistake in this book, we would be grateful if you would report this to us. Please visit www.packtpub.com/support/errata and fill in the form.

Piracy: If you come across any illegal copies of our works in any form on the internet, we would be grateful if you would provide us with the location address or website name. Please contact us at copyright@packt.com with a link to the material.

If you are interested in becoming an author: If there is a topic that you have expertise in and you are interested in either writing or contributing to a book, please visit authors.packtpub.com.

Share Your Thoughts

Once you've read *Data Wrangling with R*, we'd love to hear your thoughts! Scan the QR code below to go straight to the Amazon review page for this book and share your feedback.

https://packt.link/r/1-803-23540-3

Your review is important to us and the tech community and will help us make sure we're delivering excellent quality content.

Download a free PDF copy of this book

Thanks for purchasing this book!

Do you like to read on the go but are unable to carry your print books everywhere?

Is your eBook purchase not compatible with the device of your choice?

Don't worry, now with every Packt book you get a DRM-free PDF version of that book at no cost.

Read anywhere, any place, on any device. Search, copy, and paste code from your favorite technical books directly into your application.

The perks don't stop there, you can get exclusive access to discounts, newsletters, and great free content in your inbox daily

Follow these simple steps to get the benefits:

1. Scan the QR code or visit the link below

https://packt.link/free-ebook/9781803235400

2. Submit your proof of purchase
3. That's it! We'll send your free PDF and other benefits to your email directly

Part 1:
Load and Explore Data

This part includes the following chapters:

- *Chapter 1, Fundamentals of Data Wrangling*
- *Chapter 2, Load and Explore Datasets*
- *Chapter 3, Basic Data Visualization*

1
Fundamentals of Data Wrangling

The relationship between humans and data is age old. Knowing that our brains can capture and store only a limited amount of information, we had to create ways to keep and organize data.

The first idea of keeping and storing data goes back to 19000 BC (as stated in `https://www.thinkautomation.com/histories/the-history-of-data/`) when a bone stick is believed to have been used to count things and keep information engraved on it, serving as a tally stick. Since then, words, writing, numbers, and many other forms of data collection have been developed and evolved.

In 1663, John Graunt performed one of the first recognized data analyses, studying births and deaths by gender in the city of London, England.

In 1928, Fritz Pfleumer received the patent for magnetic tapes, a solution to store sound that enabled other researchers to create many of the storage technologies that are still used, such as hard disk drives.

Fast forward to the modern world, at the beginning of the computer age, in the 1970s, when IBM researchers Raymond Boyce and Donald Chamberlin created the **Structured Query Language (SQL)** for getting access to and modifying data held in databases. The language is still used, and, as a matter of fact, many data-wrangling concepts come from it. Concepts such as *SELECT, WHERE, GROUP BY*, and *JOIN* are heavily present in any work you want to perform with datasets. Therefore, a little knowledge of those basic commands might help you throughout this book, although it is not mandatory.

In this chapter, we will cover the following main topics:

- What is data wrangling?
- Why data wrangling?
- The key steps of data wrangling

What is data wrangling?

Data wrangling is the process of modifying, cleaning, organizing, and transforming data from one given state to another, with the objective of making it more appropriate for use in analytics and data science.

This concept is also referred to as **data munging**, and both words are related to the act of changing, manipulating, transforming, and incrementing your dataset.

I bet you've already performed data wrangling. It is a common task for all of us. Since our primary school years, we have been taught how to create a table and make counts to organize people's opinions in a dataset. If you are familiar with MS Excel or similar tools, remember all the times you have sorted, filtered, or added columns to a table, not to mention all of those lookups that you may have performed. All of that is part of the data-wrangling process. Every task performed to somehow improve the data and make it more suitable for analysis can be considered data wrangling.

As a data scientist, you will constantly be provided with different kinds of data, with the mission of transforming the dataset into insights that will, consequentially, form the basis for business decisions. Unlike a few years ago, when the majority of data was presented in a structured form such as text or tables, nowadays, data can come in many other forms, including unstructured formats such as video, audio, or even a combination of those. Thus, it becomes clear that most of the time, data will not be presented ready to work and will require some effort to get it in a ready state, sometimes more than others.

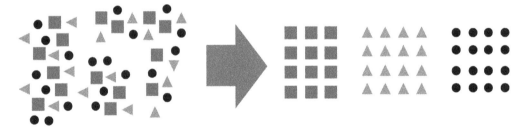

Figure 1.1 – Data before and after wrangling

Figure 1.1 is a visual representation of data wrangling. We see on the left-hand side three kinds of data points combined, and after sorting and tabulating, the data is clearer to be analyzed.

A wrangled dataset is easier to understand and to work with, creating the path to better analysis and modeling, as we shall see in the next section when we will learn why data wrangling is important to a data science project.

Why data wrangling?

Now you know what data wrangling means, and I am sure that you share the same view as me that this is a tremendously important subject – otherwise, I don't think you would be reading this book.

In statistics and data science areas, there is this frequently repeated phrase: *garbage in, garbage out*. This popular saying represents the central idea of the importance of wrangling data because it teaches us that our analysis or even our model will only be as good as the data that we present to it. You could also use the *weakest link in the chain* analogy to describe that importance, meaning that if your data is weak, the rest of the analysis could be easily broken by questions and arguments.

Let me give you a naïve example, but one that is still very precise, to illustrate my point. If we receive a dataset like in *Figure 1.2*, everything looks right at first glance. There are city names and temperatures, and it is a common format used to present data. However, for data science, this data may not be ideal for use just yet.

City / Temp (°F)	Jan	Feb	Mar	Apr	May	Jun
New York	30	35	45	50	65	74
Boston	30	32	39	49	59	68
Los Angeles	59	60	61	64	66	71

Figure 1.2 – Temperatures for cities

Notice that all the columns are referring to the same variable, which is `Temperature`. We would have trouble plotting simple graphics in R with a dataset presented as in *Figure 1.2*, as well as using the dataset for modeling.

In this case, a simple transformation of the table from wide to long format would be enough to complete the data-wrangling task.

City	Month	Temp (°F)
New York	Jan	30
New York	Feb	35
New York	Mar	45
New York	Apr	50
New York	May	65
New York	Jun	74
Boston	Jan	30
Boston	Feb	32
Boston	Mar	39
Boston	Apr	49
Boston	May	59
Boston	Jun	68
Los Angeles	Jan	59
Los Angeles	Feb	60
Los Angeles	Mar	61
Los Angeles	Apr	64
Los Angeles	May	66
Los Angeles	Jun	79

Figure 1.3 – Dataset ready for use

At first glance, *Figure 1.2* might appear to be the better-looking option. And, in fact, it is for human eyes. The presentation of the dataset in *Figure 1.2* makes it much easier for us to compare values and draw conclusions. However, we must not forget that we are dealing with computers, and machines don't process data the same way humans do. To a computer, *Figure 1.2* has seven variables: `City`, `Jan`, `Feb`, `Mar`, `Apr`, `May`, and `Jun`, while *Figure 1.3* has only three: `City`, `Month`, and `Temperature`.

Now comes the fun part; let's compare how a computer would receive both sets of data. A command to plot the temperature timeline by city for *Figure 1.2* would be as follows: *Computer, take a city and the temperatures during the months of Jan, Feb, Mar, Apr, May, and Jun in that city. Then consider each of the names of the months as a point on the x axis and the temperature associated as a point on the y axis. Plot a line for the temperature throughout the months for each of the cities.*

Figure 1.3 is much clearer to the computer. It does not need to separate anything. The dataset is ready, so look how the command would be given: *Computer, for each city, plot the month on the x axis and the temperature on the y axis.*

Much simpler, agree? That is the importance of data wrangling for Data Science.

Benefits

Performing good data wrangling will improve the overall quality of the entire analysis process. Here are the benefits:

- **Structured data**: Your data will be organized and easily understandable by other data scientists.

- **Faster results**: If the data is already in a usable state, creating plots or using it as input to an algorithm will certainly be faster.

- **Better data flow**: To be able to use the data for modeling or for a dashboard, it needs to be properly formatted and cleaned. Good data wrangling enables the data to follow to the next steps of the process, making data pipelines and automation possible.

- **Aggregation**: As we saw in the example in the previous section, the data must be in a suitable format for the computer to understand. Having well-wrangled datasets will help you to be able to aggregate them quickly for insight extraction.

- **Data quality**: Data wrangling is about transforming the data to the ready state. During this process, you will clean, aggregate, filter, and sort it accordingly, visualize the data, assess its quality, deal with outliers, and identify faulty or incomplete data.

- **Data enriching**: During wrangling, you might be able to enrich the data by creating new variables out of the original ones or joining other datasets to make your data more complete.

Every project, being related with Data Science or not, can benefit from data wrangling. As we just listed, it brings many benefits to the analysis, impacting the quality of the deliverables in the end. But to get the best from it, there are steps to follow.

The key steps of data wrangling

There are some basic steps to help data scientists and analysts to work through the data-wrangling part of the process. Naturally, once you first see a dataset, it is important to understand it, then organize, clean, enrich, and validate it before using it as input for a model.

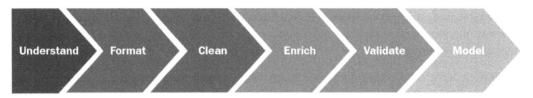

Figure 1.4 – Steps of data wrangling

1. *Understand*: The first step to take once we get our hands on new data is to understand it. Take some time to read the data dictionary, which is a document with the descriptions of the variables, if available, or talk to the owner(s) of the data to really understand what each data point represents and how they do or do not connect to your main purpose and to the business questions you are trying to answer. This will make the following steps clearer.

2. *Format*: Step two is to format or organize the data. Raw data may come unstructured or unformatted in a way that is not usable. Therefore, it is important to be familiar with the **tidy** format. Tidy data is a concept developed by Hadley Wickham in 2014 in a paper with the same name – *Tidy data* (*Tidy data. The Journal of Statistical Software, vol. 59, 2014*) – where he presents a standard method to organize and structure datasets, making the cleaning and exploration steps easier. Another benefit is facilitating the transference of the dataset between different tools that use the same format. Currently, the tidy data concept is widely accepted, so that helps you to focus on the analysis instead of munging the dataset every time you need to move it down the pipeline.

    Tidy data standardizes the way the structure of the data is linked to the semantics, in other words, how the layout is linked with the meaning of the values. More specifically, structure means the rows and columns that can be labeled. Most of the time, the columns are labeled, but the rows are not. On the other hand, every value is related to a variable and an observation. This is the data semantics. On a tidy dataset, the variable will be a column that holds all the values for an attribute, and each row associated with one observation. Take the dataset extract from *Figure 1.5* as an example. With regard to the horsepower column, we would see values such as 110, 110, 93, and 110 for four different cars. Looking at the observations level, each row is one observation, having one value for each attribute or variable, so a car could be associated with HP=110, 6 cylinders, 21 miles per gallon, and so on.

Car	Mpg	Cyl	Hp	am	Gear	carb
Mazda RX4	21	6	110	1	4	4
Mazda RX4 Wag	21	6	110	1	4	4
Datsun 710	22.8	4	93	1	4	1
Hornet 4 Drive	21.4	6	110	0	3	1

Figure 1.5 – Tidy data. Each row is one observation; each column is a variable

According to Wickham (`https://tinyurl.com/2dh75y56`), here are the three rules of tidy data:

- Every column is a variable

- Every row is an observation

- Every cell is a single value

3. *Clean*: This step is relevant to determine the overall quality of the data. There are many forms of data cleaning, such as splitting, parsing variables, handling missing values, dealing with outliers, and removing erroneous entries.

4. *Enrich*: As you work through the data-wrangling steps and become more familiar with the data, questions will arise and, sometimes, more data will be needed. That can be solved by either joining another dataset to the original one to bring new variables or creating new ones using those you have.

5. *Validate*: To validate is to make sure that the cleaning, formatting, and transformations are all in place and the data is ready for modeling or other analysis.

6. *Analysis/Model*: Once everything is complete, your dataset is now ready for use in the next phases of the project, such as the creation of a dashboard or modeling.

As with every process, we must follow steps to reach the best performance and be able to standardize our efforts and allow them to be reproduced and scaled if needed. Next, we will look at three frameworks for Data Science projects that help to make a process easy to follow and reproduce.

Frameworks in Data Science

Data Science is no different from other sciences, and it also follows some common steps. Ergo, frameworks can be designed to guide people through the process, as well as to help implement a standardized process in a company.

It is important that a Data Scientist has a holistic understanding of the flow of the data from the moment of the acquisition until the end point since the resultant business knowledge is what will support decisions.

In this section, we will take a closer look at three known frameworks that can be used for Data Science projects: **KDD**, **SEMMA**, or **CRISP-DM**. Let's get to know more about them.

KDD

KDD stands for **Knowledge Discovery in Databases**. It is a framework to extract knowledge from data in the context of large databases.

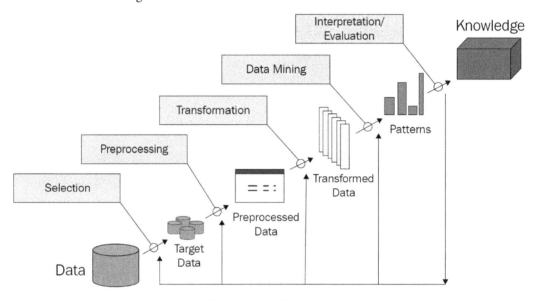

Figure 1.6 – KDD process

The process is iterative and follows these steps:

1. **Data**: Acquiring the data from a database
2. **Selection**: Creating a representative target set that is a subset of the data with selected variables or samples of interest
3. **Preprocessing**: Data cleaning and preprocessing to remove outliers and handle missing and noisy data
4. **Transformation**: Transforming and using dimensionality reduction to format the data
5. **Data Mining**: Using algorithms to analyze and search for patterns of interest (for example, classification and clustering)
6. **Interpretation/Evaluation**: Interpreting and evaluating the mined patterns

After the evaluation, if the results are not satisfactory, the process can be repeated with enhancements such as more data, a different subset, or a tweaked algorithm.

SEMMA

SEMMA stands for **Sample, Explore, Modify, Model, and Assess**. These are the steps of the process.

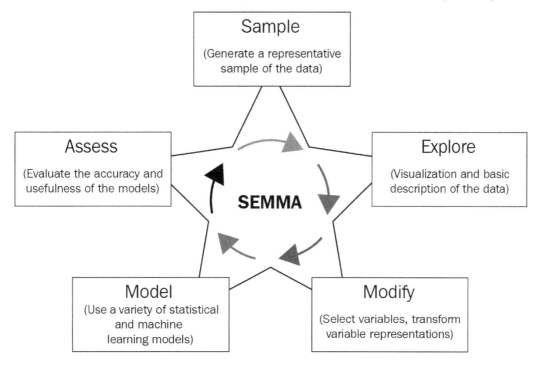

Figure 1.7 – SEMMA process

SEMMA is a cyclic process that flows more naturally with Data Science. It does not contain stages like KDD. The steps are as follows:

1. **Sample**: Based on statistics, it requires a sample large enough to be representative but small enough to be quick to work with

2. **Explore**: During this step, the goal is to understand the data and generate visualizations and descriptive statistics, looking for patterns and anomalies

3. **Modify**: Here is where data wrangling plays a more intensive role, where the transformations occur to make the data ready for modeling

4. **Model**: This step is where algorithms are used to generate estimates, predictions, or insights from the data

5. **Assess**: Evaluate the results

CRISP-DM

The acronym for this framework means **Cross-Industry Standard Process for Data Mining**. It provides the data scientist with the typical phases of the project and also an overview of the data mining life cycle.

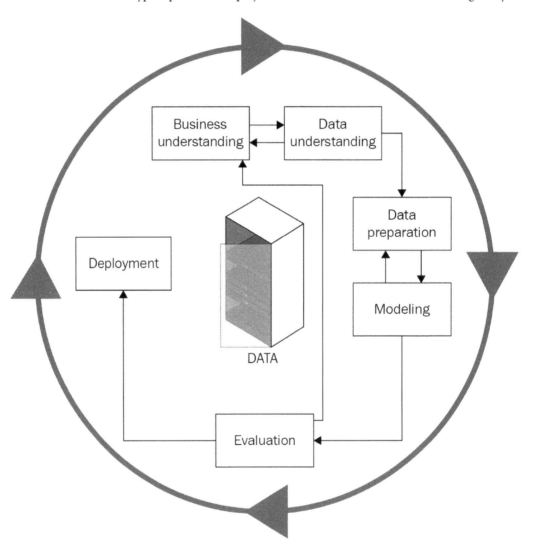

Figure 1.8 – CRISP-DM life cycle

The CRISP-DM life cycle has six phases, with the arrows indicating the dependencies between each one of them, but the key point here is that there is not a strict order to follow. The project can move back and forth during the process, making it a flexible framework. Let's go through the steps:

- **Business understanding**: Like the other two frameworks presented, it all starts with understanding the problem, the business. Understanding the business rules and specificities is often even more important than getting to the solution fast. That is because a solution may not be ideal for that kind of business. The business rules must always drive the solution.

- **Data understanding**: This involves collecting and exploring the data. Make sure the data collected is representative of the whole and get familiar with it to be able to find errors, faulty data, and missing values and to assess quality. All these tasks are part of data understanding.

- **Data preparation**: Once you are familiar with the data collected, it is time to wrangle it and prepare it for modeling.

- **Modeling**: This involves applying Data Science algorithms or performing the desired analysis on the processed data.

- **Evaluation**: This step is used to assess whether the solution is aligned with the business requirement and whether it is performing well.

- **Deployment**: In this step, the model reaches its purpose (for example, an application that predicts a group or a value, a dashboard, and so on).

These three frameworks have a lot in common if you look closer. They start with understanding the data, go over data wrangling with cleaning and transforming, then move on to the modeling phase, and end with the evaluation of the model, usually working with iterations to assess flaws and improve the results.

Summary

In this chapter, we learned a little about the history of data wrangling and became familiar with its definition. Every task performed in order to transform or enhance the data and to make it ready for analysis and modeling is what we call data wrangling or data munging.

We also discussed some topics stating the importance of wrangling data before modeling it. A model is a simplified representation of reality, and an algorithm is like a student that needs to understand that reality to give us the best answer about the subject matter. If we teach this student with bad data, we cannot expect to receive a good answer. A model is as good as its input data.

Continuing further in the chapter, we reviewed the benefits of data wrangling, proving that we can improve the quality of our data, resulting in faster results and better outcomes.

In the final sections, we reviewed the basic steps of data wrangling and learned more about three of the most commonly used frameworks for Data Science – KDD, SEMMA, and CRISP-DM. I recommend that you review more information about them to have a holistic view of the life cycle of a Data Science project.

Now, it is important to notice how these three frameworks preach the selection of a representative dataset or subset of data. A nice example is given by Aurélien Géron (*Hands-on Machine Learning with Scikit-Learn, Keras and TensorFlow, 2nd edition, (2019): 32-33*). Suppose you want to build an app to take pictures of flowers and recognize and classify them. You could go to the internet and download thousands of pictures; however, they will probably not be representative of the kind of pictures that your model will receive from the app users. Ergo, the model could underperform. This example is relevant to illustrate the garbage in, garbage out idea. That is, if you don't explore and understand your data thoroughly, you won't know whether it is good enough for modeling.

The frameworks can lead the way, like a map, to explore, understand, and wrangle the data and to make it ready for modeling, decreasing the risk of having a frustrating outcome.

In the next chapter, let's get our hands on R and start coding.

Exercises

1.   What is data wrangling?

2.   Why is data wrangling important?

3.   What are the steps for data wrangling?

4.   List three Data Science frameworks.

Further reading

- Hadley Wickham – *Tidy Data*: https://tinyurl.com/2dh75y56

- What is data wrangling?: https://tinyurl.com/2p93juzn

- Data Science methodologies: https://tinyurl.com/2ucxdcch

2
Loading and Exploring Datasets

Every data exploration begins with data, quite obviously. Thus, it is important for us to know how to load datasets to RStudio before we get to work. In this chapter, we will learn the different ways to load data to an RStudio session. We will begin by importing some sample datasets that come with some preinstalled libraries from R, then move on to reading data from Comma-Separated Values (CSV) files, which turns out to be one of the most used file types in Data Science, given its compatibility with many other programs and data readers.

We will also learn about the basic differences between a Data Frame and a Tibble, followed by a section where we will learn the basics of Web Scraping, which is another good way to acquire data. Later in the chapter, we will learn how to save our data to our local machine and paint a picture of a good workflow for data exploration.

We will cover the following main topics:

- How to load files to RStudio
- Tibble versus Data Frame
- Basic Web Scraping
- Saving files
- A workflow for data exploration

Technical requirements

To make yourself ready for this chapter, make sure to install the following libraries by using the `install.packages("library_name")` command:

- `datasets`, `lubridate`, `mice`, `corrplot`, `tidyverse`, `httr`, `jsonlite`, and `rvest`

Let's get to work.

How to load files to RStudio

RStudio, just like other programming languages, provides many preinstalled libraries and contents for demonstration and educational purposes. Therefore, the easiest way to load data and start working in R is to use the `utils` library. This library does not usually need to be imported using the `library(utils)` command. You just need to write the following command:

```
data("Orange")
```

You might see that the dataset won't load promptly and may notice a message saying **<Promise>** on the **Environment** tab. If that happens, just click on the **<Promise>** value, and it will force the data to be loaded.

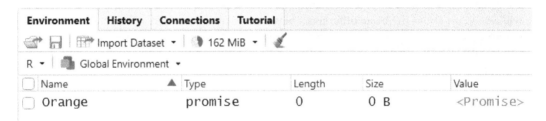

Figure 2.1 – Orange dataset in <Promise> mode

And as a result, you will see (*Figure 2.2*) the toy dataset with information about the growth of orange trees.

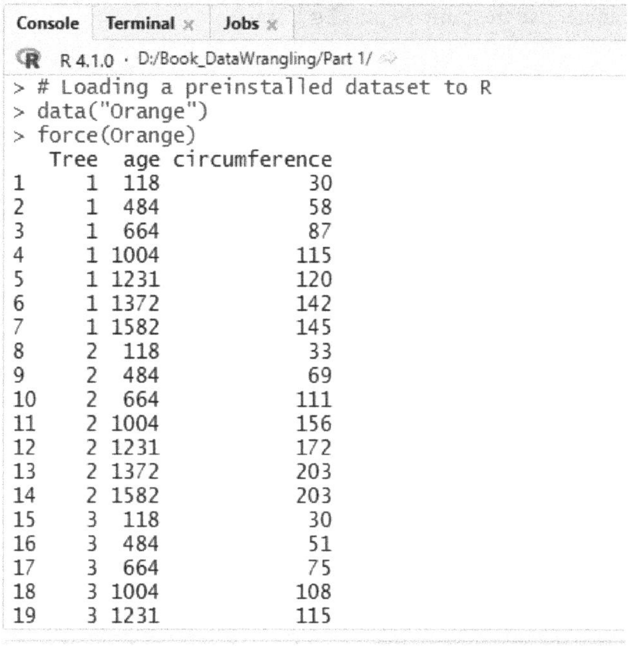

```
Console   Terminal ×   Jobs ×
R  R 4.1.0 · D:/Book_DataWrangling/Part 1/
> # Loading a preinstalled dataset to R
> data("Orange")
> force(Orange)
   Tree  age circumference
1     1  118            30
2     1  484            58
3     1  664            87
4     1 1004           115
5     1 1231           120
6     1 1372           142
7     1 1582           145
8     2  118            33
9     2  484            69
10    2  664           111
11    2 1004           156
12    2 1231           172
13    2 1372           203
14    2 1582           203
15    3  118            30
16    3  484            51
17    3  664            75
18    3 1004           108
19    3 1231           115
```

Figure 2.2 – Orange dataset displayed on the Console screen

You can see in *Figure 2.3* that the data is now loaded under the variable name of **Orange**, the type is **nfnGroupedData** – a kind of Data Frame – and it has 35 rows and 3 columns.

So RStudio does a really good job of providing information about the objects you are working with in the **Environment** tab. From there, we can say we have already started our analysis by learning about and understanding the basic metadata.

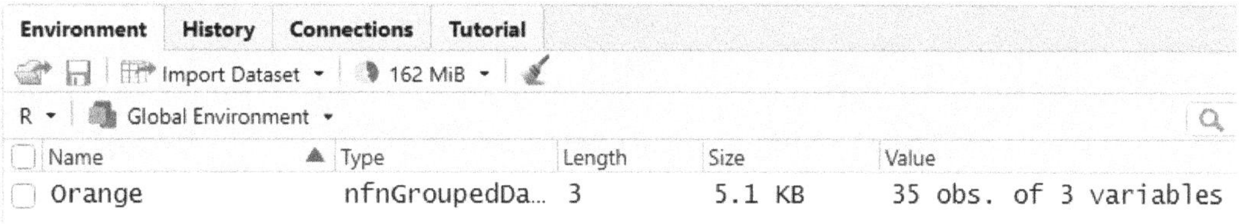

```
Environment   History   Connections   Tutorial
      Import Dataset ▾    162 MiB ▾
R ▾    Global Environment ▾                                          Q
  Name            ▲ Type            Length   Size      Value
  Orange            nfnGroupedDa... 3        5.1 KB    35 obs. of 3 variables
```

Figure 2.3 – Metadata of the Orange dataset

Loading a CSV file to R

CSV is possibly one of the most popular file formats for rectangular datasets (rows x columns). This is because CSV files are basically textual files with a minimum separation of columns by a comma, which makes them extremely compatible with many programs and file readers, ergo, easily shareable.

To load a CSV in R, you can use the built-in `utils` library:

```
# Setup the file path
path = "C:/Users/user_name/Desktop/oranges.csv"
# Load the file
df <- read.csv(path)
# View the dataset
View(df)
```

In the first line of code, we pointed out where R should look for the `Oranges` dataset on our local machine. Then we assigned it to the `df` variable and, finally, displayed it with R's viewer. You will see in *Figure 2.4* that the result is the very same as in *Figure 2.2*.

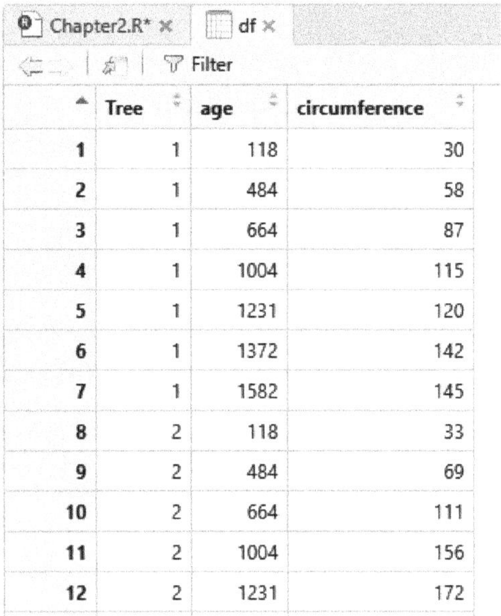

	Tree	age	circumference
1	1	118	30
2	1	484	58
3	1	664	87
4	1	1004	115
5	1	1231	120
6	1	1372	142
7	1	1582	145
8	2	118	33
9	2	484	69
10	2	664	111
11	2	1004	156
12	2	1231	172

Figure 2.4 – The Oranges dataset loaded using CSV

You can also use `read.csv2()` if you have a CSV file separated by semicolons (;), although that is also customizable using the `sep` parameter from the `read.csv()` command. Other interesting parameters that can be used are listed as follows. For the full list of options, just explore the help supplied for the function by clicking with the mouse cursor on the function name and then pressing *F1* on your keyboard:

- `header`: This is set to TRUE by default, indicating that the first row is the header. But if you have a dataset without a header, you can set it to FALSE and add the column names using the `col.names` parameter.

sep: By default, it is set to either a comma or semicolon, depending on the function you choose, but you can change it to whatever you need, such as the pipe sign, which is also common (|).

col.names: You can pass a custom vector with column names for your dataset.

row.names: You can pass a custom vector with row names for your dataset.

CSV files are so popular that you will find a couple of different functions to load them. Next, we will learn about another way to load a CSV file.

Loading a CSV file using tidyverse's readr

Another way to load a CSV to R is using the readr library. This library is part of the **tidyverse** package, so you don't require both if the other is already loaded.

To import a CSV file using this library, you use pretty much the same steps you followed with the utils package:

```
# Load the tidyverse library that contains readr
library(tidyverse)
# Setup the file path
path = "C:/Users/user_name/Desktop/oranges.csv"
# Load the file
df <-    read_csv(path)
```

This is what the **Console** screen displays after we hit *Ctrl + Enter* on the keyboard or select the code snippet and click on the **Run** button:

```
Console   Terminal ×   Jobs ×
R   R 4.1.0 · D:/Book_DataWrangling/Part 1/
> df <-   read_csv(path)
Rows: 35 Columns: 3

── Column specification ──────────────────────────────────────────────────────
Delimiter: ","
dbl (3): Tree, age, circumference

i Use `spec()` to retrieve the full column specification for this data.
i Specify the column types or set `show_col_types = FALSE` to quiet this message.
> |
```

Figure 2.5 – CSV file loaded using readr in tibble format

What is interesting about readr is that it uses the Tibble format, so you may notice some tweaks, such as the delimiter used and the data types of the columns, along with the spec() function. If we enter that function on the **Console** screen, here's what it shows:

```
Console   Terminal ×   Jobs ×
R   R 4.1.0 · D:/Book_DataWrangling/Part 1/
> spec(df)
cols(
  Tree = col_double(),
  age = col_double(),
  circumference = col_double()
)
>
```

Figure 2.6 – Using spec(df); it shows information about the data types

Moving on, next we will see how to load a CSV file using **data.table**.

Loading a CSV file using data.table's fread

The last option covered is the `fread()` function from the `data.table` library. Using it is not very different from the other two options. Set up a path where the CSV file is and use the function, as demonstrated here:

```
# Load the library
library(data.table)
# setup the file path
path = 'C:/Users/.../oranges.'sv'
# Load the file
df <-    data.table::fread(path)
```

Notice that, in this case, the object type will be the `data.table` format, which is different from a data frame or tibble. To convert the file into the `data.frame` type, use `as.data.frame(df)`.

Comparing functions to load CSV files

You might be asking yourself, which one is the best function to load a file into RStudio? Well, that question can be answered in different ways, as it depends on the size of the file and the library to be used for the project. For our purpose here, let's consider the time to upload a file as our main metric of comparison.

After creating a test dataset with 735,000 observations and recording the time elapsed to load the CSV, the graphic from *Figure 2.7* shows that the `fread()` function from the `data.table` library is the fastest option, while the `read.csv()` command, the built-in function from R, is the slowest. Tidyverse's `read_csv()` option is the mid-point option since it is fairly fast and it is the native function from the package that carries most of the data-wrangling tools.

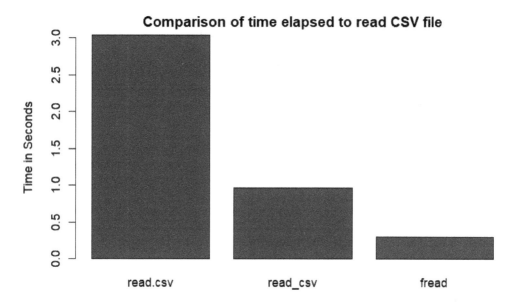

Figure 2.7 – Time comparison between the three functions to load CSV files

Customizing data while importing

Loading the dataset is the beginning of the data-wrangling process and can be used to start manipulating the data while pulling it into an R session. Customizations include selecting a subset of columns or a maximum number of rows to read, treating missing data, using custom column names, and trimming white spaces.

In order to complete the following exercises and demonstrate the parameters properly, I have modified the Oranges dataset a bit, adding a duplicated column name (circumference) with random values and a couple of missing data points. Let's look at how those parameters can be used to fix the problems:

```
# Load a file with only the original columns and 10 rows
df <- read_csv(path, col_select = c(1, 2, 3), n_max = 10)
# Show the first 3 rows
df %>% head(3)
# A tibble: 6 x 3
     Tree      age circumference...3
    <dbl> <dbl>                      <dbl>
1        1      118                        30
2        1      484                        58
3        1      664                        87
```

The `col_select` parameter will let you add the number or the names of the desired columns to load. In our case, if we use the names, R will throw an error saying that there are duplicate names, and it can't understand which column circumference it should load. However, `df <- read_csv(path, col_select =c("Tree", "age"))` would work fine. The `n_max` argument says how many rows we want to load. It can be useful if you are working with a really large dataset and you want to load just a few rows at first to become familiar with the dataset, being able to work at a faster speed using less memory.

Notice that I have used the `%>%` sign in the code. This is called the pipe operator, commonly used in `tidyverse` coding with the purpose of gathering multiple functions to be applied over the same object – `df` in our case – therefore reducing repetition. It follows the same logic as the dot notation in Python, for those who are familiar with that language.

Now, let's move on to missing data. The missing values will not always be blank cells. Sometimes, the team who owns the data can use their own convention. Imagine that you have a dataset with blanks and **No Record** entries as missing data. See row 9 of *Figure 2.8*, for example.

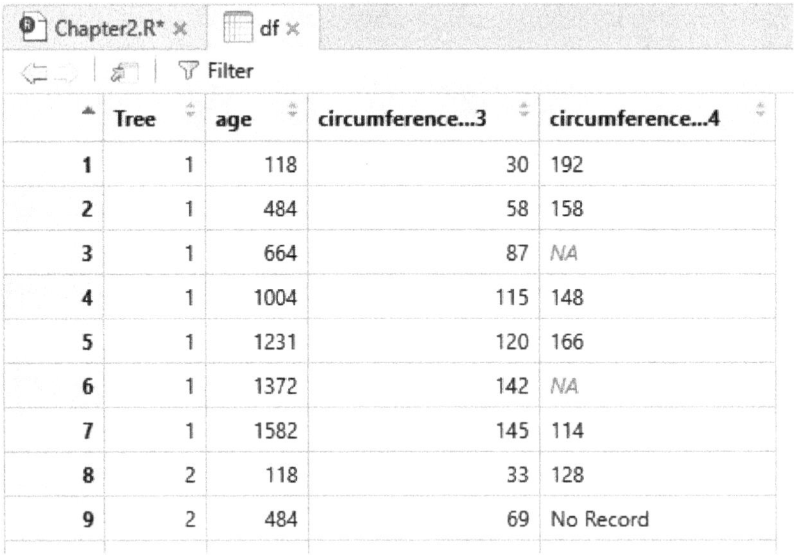

	Tree	age	circumference...3	circumference...4
1	1	118	30	192
2	1	484	58	158
3	1	664	87	NA
4	1	1004	115	148
5	1	1231	120	166
6	1	1372	142	NA
7	1	1582	145	114
8	2	118	33	128
9	2	484	69	No Record

Figure 2.8 – Missing data

We can deal with that using the `na` parameter, passing a vector with the entries we want to be treated as NA. The function will treat blank entries as NA by default, but anything else should be passed to the `read_csv` function, just like the following code snippet, where we added `No Record` to the list of missing data:

```
# Listing entries to consider NA
df <- read_csv(path, na = c("","No Record") )
```

	Tree	age	circumference...3	circumference...4
1	1	118	30	192
2	1	484	58	158
3	1	664	87	NA
4	1	1004	115	148
5	1	1231	120	166
6	1	1372	142	NA
7	1	1582	145	114
8	2	118	33	128
9	2	484	69	NA

Figure 2.9 – Missing data listed is all converted to NA

To close this section, I would like to just point out the `trim_ws` parameter, which by default trims white spaces from the beginning and end of the data, and `name_repair`, which ensures the default value is *unique*, by changing the names of repeated column names to make them distinct. In *Figure 2.9*, the duplicated circumference column name is made distinct with the addition of a number.

Tibbles versus Data Frames

A tibble is a modern data frame. With the evolution of the R language, some features that used to work well for data frames are not very useful now, so Hadley Wickham – the creator of Tidy Data and `tidyverse` – created a new data type that better fitted his way of working with datasets and dropped the old features, while adding other fancy ones, to create this new data type named Tibble.

Let's refer to the documentation:

> *they do less (i.e. they don't change variable names or types, and don't do partial matching) and complain more (e.g. when a variable does not exist). This forces you to confront problems earlier, typically leading to cleaner, more expressive code.*
> `(https://tinyurl.com/43un2f3z)`

The main differences between Tibbles and data frames are as follows:

- They don't change input variable types, such as strings to factors, by default.
- They can have lists as columns.

- They can have non-standard variable names – such as names with spaces.

- They only recycle vectors with a length of 1.

- They print an abbreviated description of the column types when using the head() function.

- They always return another Tibble when slicing. Data frames sometimes return vectors. That is the reason why some libraries don't deal well with Tibbles. If you face an error requesting a data.frame object as input, just use as.data.frame() to convert your object.

As it was developed to follow the Tidy Data concept, the Tibble format works much better with the tidyverse package. When you use a data frame as input to a function from the tidyverse package, such as dplyr::group_by(), you will see an object, tbl_df, as the output type.

When printing a Tibble on the **Console** screen, you can choose to see all the columns by setting the width to infinite, as in the following code:

```
# Print all the columns using tibble
data("world_bank_pop")
world_bank_pop%>% print(width=Inf)
```

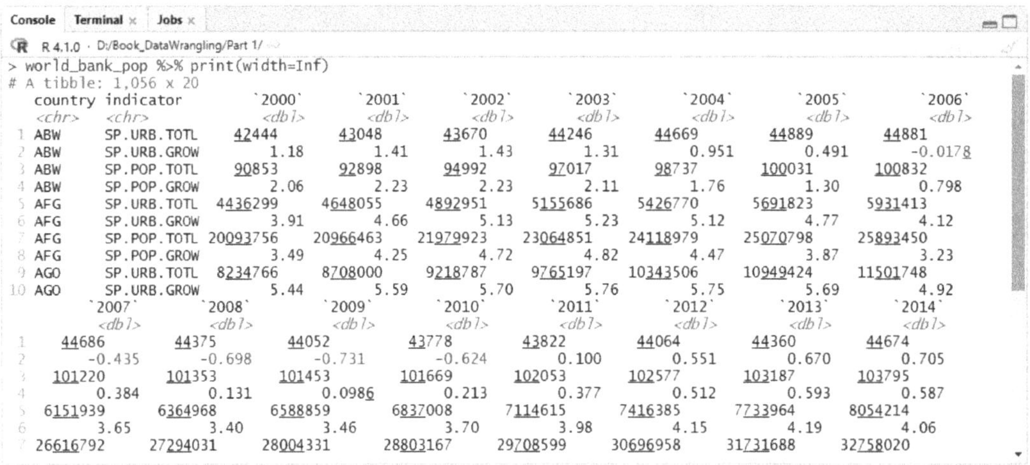

Figure 2.10 – Print all the variables using an infinite width for Tibbles

Although it is possible to see all the columns, it is not a user-friendly format, especially if you are dealing with a dataset with many variables, as seen in *Figure 2.10*. I suggest the use of RStudio's viewer for this, accessible by clicking on the data frame's name on the **Environment** tab or using the View(df) code.

Next, we will learn how to save files in CSV format using RStudio.

Saving files

During data exploration or modeling, it becomes crucial to save our work so we can continue from a certain point without the need to rerun the code all over again. This becomes especially interesting when the dataset you are dealing with is particularly large. Large datasets can take much longer to run some operations, so you want to save your clean data in order to pick up from there the next time you revisit that work.

Saving files in R is simple. The `utils` library takes care of that. The following code saves the content of the `df` variable – which is the **Monthly Treasury Statement** (**MTS**) pulled from the API – to a CSV file in the same working directory of the script or `Rproj` file:

```
# Save a variable to csv
write.csv(df, "Monthly_Treasury_Statement.csv", row.names =
FALSE)
```

The `row.names` parameter is used to omit the index column. You can do the same task using the `readr` version too, `write_csv(df, "file_name.csv")`.

Furthermore, for a data scientist working as part of a team, file sharing is a common task. There are certainly better ways to control versions and share results, but for multidisciplinary teams composed of non-technical people, the use of CSV files is still important, enforcing the need for ways to save files from RStudio in that format.

We have learned how to load data to RStudio. Following this, it is time to look at how to explore the dataset you have in your hands. This will be our first time exploring data in this book; therefore, we will start with a basic view of the process. In *Chapter 9*, we will go deeper into data exploration.

A workflow for data exploration

Now that you are familiar with the different ways to acquire and load data into Rstudio, let's go over a basic workflow that I regularly use for data exploration. Naturally, the steps presented here are flexible and should be understood as a guide to begin understanding the dataset. It can and should be changed to adapt to your project's needs.

When you start a data exploration, it is important to have in mind your final goal. What problem are you trying to solve? Then, you look to understand the variables, look for errors and missing data, understand the distributions, and create a couple of visualizations that will help you to extract good insights to help you along the way. Let's explore the steps that can be performed:

1. **Load and view**: Every Data Science project starts with data. Load a dataset to RStudio and take a first look at it, making sure the data types are correctly inferred and that the dataset is completely loaded without errors.

2. **Descriptive statistics**: Display the descriptive statistics of the numeric variables. This helps us to see how the data is spread and gives us a sense of possibly faulty or missing data.

3. **Missing values**: Look for missing values, such as NA or NULL, and give them proper treatment.

4. **Data distributions**: After the dataset has been cleaned of missing data, look at the distributions of the variables and understand how they are presented compared to the average.

5. **Visualizations**: An image is worth a thousand words, as it is said. Creating visualizations helps us to get a feeling of the best variables for modeling.

Let's look at these steps in detail.

Loading and viewing

It is our job as data scientists to explore data, extract value from it, and answer business questions using it. Ergo, our workflow begins with loading a dataset and displaying it to get familiar with the variables available and to connect to the problem to be solved.

A good **exploratory data analysis** (**EDA**) begins with questions. After we load and look at the variables, we must start formulating questions that will lead to the exploration and understanding of the data, as well as the formulation of new questions. When you use the question-and-answer methodology, it helps you to focus on specific parts of the dataset and create a deeper connection with that information.

Suppose that our client is a real estate agency that wants to predict the values of houses in Texas. From their database, they collect a dataset and assign you the mission of analyzing the data and extracting the best insights for a possible future model that will predict median sales prices.

For this exercise, we will use the `txhousing` dataset from the `datasets` library. It provides information about the housing market in Texas provided by the TAMU real estate center (`https://www.recenter.tamu.edu/`). The variables are as follows:

- **City**: The name of the city

- **Year**: YYYY format

- **Month**: MM format

- **Sales**: The number of sales on that date

- **Volume**: The total value of the sales

- **Median**: The median sale price

- **Listings**: The total active listings

- **Inventory**: The *months inventory* – the amount of time it would take to sell all current listings at the current pace of sales

- **Date**: The date formatted as float numbers

After loading the dataset to be used with `data("txhousing")`, the next step is to look at it. We will use RStudio's viewer, which can be done with the following code:

```
# Visualize the dataset
View(txhousing)
```

	city	year	month	sales	volume	median	listings	inventory	date
1	Abilene	2000	1	72	5380000	71400	701	6.3	2000.000
2	Abilene	2000	2	98	6505000	58700	746	6.6	2000.083
3	Abilene	2000	3	130	9285000	58100	784	6.8	2000.167
4	Abilene	2000	4	98	9730000	68600	785	6.9	2000.250
5	Abilene	2000	5	141	10590000	67300	794	6.8	2000.333
6	Abilene	2000	6	156	13910000	66900	780	6.6	2000.417
7	Abilene	2000	7	152	12635000	73500	742	6.2	2000.500
8	Abilene	2000	8	131	10710000	75000	765	6.4	2000.583
9	Abilene	2000	9	104	7615000	64500	771	6.5	2000.667
10	Abilene	2000	10	101	7040000	59300	764	6.6	2000.750
11	Abilene	2000	11	100	7890000	70900	721	6.2	2000.833
12	Abilene	2000	12	92	7285000	65000	658	5.7	2000.917
13	Abilene	2001	1	75	5730000	64500	779	6.8	2001.000

Figure 2.11 – Texas housing dataset in RStudio viewer

Let's look at the data types of each variable. In Object-Oriented Programming languages, it is important to do that because the type of data will determine which operations you can perform, or the methods and attributes that you can use with that object:

```
glimpse(txhousing)
```

The preceding code will display the following results (*Figure 2.12*):

```
Rows: 8,602
Columns: 9
$ city      <chr> "Abilene", "Abilene", "Abilene", "Abilene", "Abilene", "Abilen~
$ year      <int> 2000, 2000, 2000, 2000, 2000, 2000, 2000, 2000, 2000, 2000, 20~
$ month     <int> 1, 2, 3, 4, 5, 6, 7, 8, 9, 10, 11, 12, 1, 2, 3, 4, 5, 6, 7, 8,~
$ sales     <dbl> 72, 98, 130, 98, 141, 156, 152, 131, 104, 101, 100, 92, 75, 11~
$ volume    <dbl> 5380000, 6505000, 9285000, 9730000, 10590000, 13910000, 126350~
$ median    <dbl> 71400, 58700, 58100, 68600, 67300, 66900, 73500, 75000, 64500,~
$ listings  <dbl> 701, 746, 784, 785, 794, 780, 742, 765, 771, 764, 721, 658, 77~
$ inventory <dbl> 6.3, 6.6, 6.8, 6.9, 6.8, 6.6, 6.2, 6.4, 6.5, 6.6, 6.2, 5.7, 6.~
$ date      <dbl> 2000.000, 2000.083, 2000.167, 2000.250, 2000.333, 2000.417, 20~
```

Figure 2.12 – Variables and data types of the dataset

R did a pretty good job of inferring the data types, but they will need some adjustments. The date variable should be the datetime type, while year and month could be either numeric or even a factor since we're not using those as numbers but more like categories. Likewise, city should be converted into a factor, and you will see why in a bit. Let's perform the changes using the following code:

```
# Let's create a copy of the dataset to preserve the original
txhouses <- txhousing
# Adjusting data types
txhouses$date <- date_decimal(txhouses$date, tz="GMT")
txhouses$city <- as.factor(txhouses$city)
glimpse(txhouses)
```

We used date_decimal() to convert a decimal input into a datetime object and as.factor() to make text a category. Now, let's check whether it was performed correctly:

```
Rows: 8,602
Columns: 9
$ city      <fct> Abilene, Abilene, Abilene, Abilene, Abilene, Abilene, Abilene, Abilene, Abilene, A~
$ year      <int> 2000, 2000, 2000, 2000, 2000, 2000, 2000, 2000, 2000, 2000, 2000, 2000, 2001, 2001~
$ month     <int> 1, 2, 3, 4, 5, 6, 7, 8, 9, 10, 11, 12, 1, 2, 3, 4, 5, 6, 7, 8, 9, 10, 11, 12, 1, 2~
$ sales     <dbl> 72, 98, 130, 98, 141, 156, 152, 131, 104, 101, 100, 92, 75, 112, 118, 105, 150, 13~
$ volume    <dbl> 5380000, 6505000, 9285000, 9730000, 10590000, 13910000, 12635000, 10710000, 761500~
$ median    <dbl> 71400, 58700, 58100, 68600, 67300, 66900, 73500, 75000, 64500, 59300, 70900, 65000~
$ listings  <dbl> 701, 746, 784, 785, 794, 780, 742, 765, 771, 764, 721, 658, 779, 700, 738, 810, 77~
$ inventory <dbl> 6.3, 6.6, 6.8, 6.9, 6.8, 6.6, 6.2, 6.4, 6.5, 6.6, 6.2, 5.7, 6.8, 6.0, 6.4, 7.0, 6.~
$ date      <dttm> 2000-01-01 00:00:00, 2000-01-31 11:59:59, 2000-03-02 00:00:00, 2000-04-01 12:00:0~
```

Figure 2.13 – Adjusted variables and data types

Our job in this step is done. We now have a better understanding of the variables.

Descriptive statistics

Descriptive statistics help us to get a feeling for how our data is distributed. With a single line of code, we can take so much information from the data and start raising some good questions. Take the following line of code:

```
# Descriptive statistics
summary(txhouses)
```

The following information is displayed:

```
      city              year             month             sales             volume
 Abilene  : 187    Min.   :2000    Min.   : 1.000    Min.   :   6.0    Min.   :8.350e+05
 Amarillo : 187    1st Qu.:2003    1st Qu.: 3.000    1st Qu.:  86.0    1st Qu.:1.084e+07
 Arlington: 187    Median :2007    Median : 6.000    Median :  169.0   Median :2.299e+07
 Austin   : 187    Mean   :2007    Mean   : 6.406    Mean   :  549.6   Mean   :1.069e+08
 Bay Area : 187    3rd Qu.:2011    3rd Qu.: 9.000    3rd Qu.:  467.0   3rd Qu.:7.512e+07
 Beaumont : 187    Max.   :2015    Max.   :12.000    Max.   :8945.0    Max.   :2.568e+09
 (Other)  :7480                                      NA's   :568       NA's   :568
     median           listings          inventory           date
 Min.   : 50000   Min.   :    0    Min.   : 0.000    Min.   :2000-01-01 00:00:00
 1st Qu.:100000   1st Qu.:  682    1st Qu.: 4.900    1st Qu.:2003-11-01 04:00:00
 Median :123800   Median : 1283    Median : 6.200    Median :2007-10-01 18:00:00
 Mean   :128131   Mean   : 3217    Mean   : 7.175    Mean   :2007-10-02 01:42:01
 3rd Qu.:150000   3rd Qu.: 2954    3rd Qu.: 8.150    3rd Qu.:2011-09-01 08:00:00
 Max.   :304200   Max.   :43107    Max.   :55.900    Max.   :2015-07-02 12:00:00
 NA's   :616      NA's   :1424     NA's   :1467
```

Figure 2.14 – Descriptive statistics

Once I call the `summary()` function, native in R, I can see all the important descriptive statistics for each variable. For `city` – and here is why we converted it into the factor type – we see the most frequent appearances, something that does not happen if you keep it as a character. One curious fact is that the cities appear in alphabetical order, and they have the same frequency. This can provide a question for further exploration: *do all the cities appear the same number of times – 187 times each?*

If we look at the data again, we will see that the city of Abilene has entries for each month in 2000, 2001, 2002, up to 2015. And that repeats for every city. That is one indication that the number 187 is equal for each city. Now, let's take the total number of rows and divide it by the number of cities and see what we get:

```
nrow(txhouses) / length(unique(txhouses$city))
[1] 187
```

The value is `187`, and it answers our first exploration question. All the cities appear the same number of times.

Other insights made from here are as follows:

- The data is spread between 2000 and 2015, with the mean and median right in the middle, indicating that there is no predominance of data from any specific year. There's probably the same data for each year. We can draw a similar conclusion for months.

- On average, there are 549 sales per period, but the data is widely spread, given that the minimum and maximum values are very far from the mean and median.

- The volume is also apparently spread out, but the scientific notation makes it harder to be sure, so that will be verified during the visualizations we will carry out.

- The median sales price is apparently well spread around the mean, with the mean and median close to 123k and 128k, respectively.

- The listings also look very spread out but the inventory is tighter, with the time taken to sell all the houses floating around 6 to 7 months.

- There are missing values in the `sales`, `volume`, `median`, `listings`, and `inventory` variables.

Now, look at the amount of information that I was able to capture before I even calculated a mean or plotted a distribution. Can you understand the power of this step?

I believe you can now.

Missing values

A missing value is a data value that is not available for a given variable in an observation. In plain English, it is when you look at a row of the dataset, and there is one value not available for a column. If you think it through, even missing data brings some information with it. Sometimes it can be a record that could not be captured, or it can indicate a failure in the process that is preventing data points from being registered for a specific variable.

In general, there are two ways to deal with missing data. The first one is to drop them all. A widely used rule of thumb says that you can drop all NAs if they make up to 5% of the total observations. This makes a lot of sense since it would be a waste of time trying to handle them for such a small amount that won't affect the modeling that much in most cases. Another argument in favour of this, in this case, is the bias. Every transformation will be – even if only slightly – biased by the data scientist's decision as to what they will replace the missing values with.

This is one possible way to drop the NAs in R:

```
# Drop NAs
txhouse_no_na <- txhouses %>% na.omit()
```

You will see the following results:

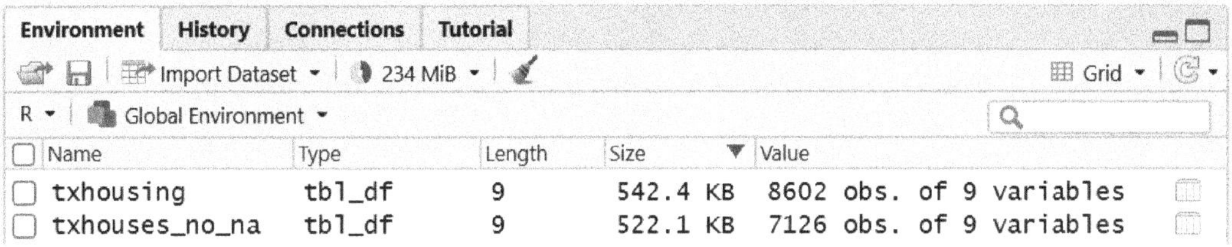

Figure 2.15 – Dataset after NAs are dropped

In this case, we can see that we have dropped 1,476 observations, or roughly 17% of the data, which is way beyond our cutoff rule of 5%. We must handle the missing data another way.

That is where we get to the second way of handling missing values, which is to replace them with some other value. There are approaches ranging from simply replacing the values with the mean, median, or most frequent value of the column to more complex approaches such as using interpolation or machine learning algorithms to deal with it.

In our exercise, we will combine both methods. If you look at *Figure 2.16*, from rows 12 to 23, the observations for Brazoria County for 2003 have only the city name, year, month, and date. Nothing else. This is the kind of missing data that must be dropped, just because there is not a good way to recover it. Any interpolation or machine learning method here would basically create new data, which could negatively influence our model.

	city	year	month	sales	volume	median	listings	inventory	date
8	Bay Area	2001	9	402	54767306	115400	NA	NA	2001-09-01 08:00:00
9	Brazoria County	2001	3	95	8560000	77100	NA	NA	2001-03-02 20:00:00
10	Brazoria County	2001	6	115	13105000	89600	NA	NA	2001-06-02 02:00:00
11	Brazoria County	2001	10	NA	NA	NA	NA	NA	2001-10-01 18:00:00
12	Brazoria County	2003	1	NA	NA	NA	NA	NA	2003-01-01 00:00:00
13	Brazoria County	2003	2	NA	NA	NA	NA	NA	2003-01-31 09:59:59
14	Brazoria County	2003	3	NA	NA	NA	NA	NA	2003-03-02 20:00:00
15	Brazoria County	2003	4	NA	NA	NA	NA	NA	2003-04-02 06:00:00
16	Brazoria County	2003	5	NA	NA	NA	NA	NA	2003-05-02 15:59:59
17	Brazoria County	2003	6	NA	NA	NA	NA	NA	2003-06-02 02:00:00
18	Brazoria County	2003	7	NA	NA	NA	NA	NA	2003-07-02 12:00:00
19	Brazoria County	2003	8	NA	NA	NA	NA	NA	2003-08-01 21:59:59
20	Brazoria County	2003	9	NA	NA	NA	NA	NA	2003-09-01 08:00:00
21	Brazoria County	2003	10	NA	NA	NA	NA	NA	2003-10-01 18:00:00
22	Brazoria County	2003	11	NA	NA	NA	NA	NA	2003-11-01 03:59:59
23	Brazoria County	2003	12	NA	NA	NA	NA	NA	2003-12-01 14:00:00

Figure 2.16 – Missing data that can't be recovered

So, let's drop observations of this kind. Using the `which()` function will return the position or the index of the value, which satisfies the given condition in a vector. So, in this case, we are interested in knowing what the indexes are of the rows that have five NA cells. In the sequence, we use a slicing notation to return all the rows except those from the index. We do that with the minus sign, `[-idx,]`, in square brackets:

```
# Find the row numbers with 5 NAs
idx <- which( rowSums(is.na(txhouses)) == 5 )
# Filter those rows out
txhouses <- txhouses[-idx,]
```

If we look at the `listings` and `inventory` variables, we will also see many NAs. So, let's check the proportion of NAs versus non-NAs for both of those columns. If that is too high, such as 50% or more, there is no point in keeping them. The `prop.table()` function returns the proportion of each value as compared to all values, and we want to calculate that over the table count passed as an argument to the function:

```
# Proportion listings
print("Proportions to `listings`")
prop.table( table(is.na(txhouses$listings)))
# Proportion invetory
writeLines("-------------------------------")
writeLines("\nProportions to `inventory`")
prop.table( table(is.na(txhouses$inventory)))
```

```
[1] "Proportions to `listings`"
        FALSE              TRUE
0.8931193 0.1068807
-------------------------------
Proportions to `inventory`
        FALSE              TRUE
0.8877691 0.1122309
```

Both variables are floating around 11% missing values. In this case, we will have to handle the NAs with an imputation of data because if we drop both variables, we will let 89% of the data go with them.

Similarly, we could use the same code to check the other variables and see that they have less than circa 0.5% missing values. For those, let's just use the median value to impute, slicing the dataset only for missing values and adding the median calculation by variable:

```
# Impute median value
txhouses$sales[is.na(txhouses$sales)] <- median(txhouses$sales,
na.rm=T)
txhouses$volume[is.na(txhouses$volume)] <-
median(txhouses$volume, na.rm=T)
txhouses$median[is.na(txhouses$median)] <-
median(txhouses$median, na.rm=T)
```

Finally, let's load `library(mice)` to impute the data for the `listings` and `inventory` variables. This library uses different algorithms for the prediction of data, such as regression, logistic regression, polynomial regression, and decision trees. These are fancy ways to handle NAs. The following is the code to perform the imputation. The function to create the imputer is `mice()`, which receives the columns to be changed plus a seed, which is optional, to always generate the same result instead of random numbers for every run. To create the replacement columns with the created data, we use `complete()`. To add that to the dataset itself, the `mutate()` function is used with a column of 1 for `listings` and a column of 2 for `inventory`:

```
# Use mice inputer
impute <- mice(data.frame(txhouses[,7:8]), seed = 123)
impute_data <- complete(impute, 1)
# Replace the columns with the imputed ones
txhouses_clean <- txhouses %>%
    mutate(listings = impute_data[,1],
                    inventory = impute_data[,2])
```

We can see that the algorithm did a nice job. In *Figure 2.17*, I compare both distributions, before and after the imputation, as the lines almost completely overlap.

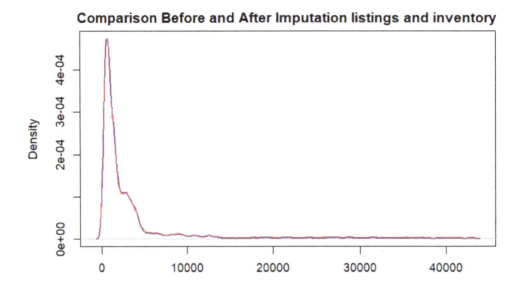

Figure 2.17 – Comparison of the listings and inventory variables
before and after imputation with the mice library

The missing data is now gone:

```
# Checking missing data
sum( is.na(txhouses_clean) )
```

```
[1] 0
```

With the dataset clean of missing values, we can start looking at data distributions. It is important to know where the gross of the data is concentrated and to detect those points that lie outside of the curve, the ones called *outliers*.

Data distributions

Every variable is a collection of data points about some matter. If we measure the number of people on a subway train at 10 a.m. and again at 5 p.m., we will most certainly get two different numbers. If we repeat that process hundreds of times, the numbers will keep changing. That is what we call variation. Every variation will have some pattern, such as a higher concentration of low values, a higher concentration of high values, or no defined pattern, such as a random collection of numbers.

Knowing this supports our point that we should look at the distributions of our variables to understand what those variation patterns are. For categorical data, if we look at a response variable – the one we will try to predict – we see that, there is much more of group A than group B, and it becomes natural to conclude that our model will be biased to classify more As than Bs. Similarly, in regression models, if there is a high concentration of low values, we know that high values can be more difficult to predict.

To check the data distributions, the best types of graphics are **histograms**, **boxplots**, and **density plots**. Histograms tell us where the data is more concentrated, revealing its format: symmetrical, left-skewed with a longer tail to the left, or right-skewed with a longer tail to the right. Boxplots bring interesting information, such as the median, quartiles, and outliers. Density plots help us see the format of our distribution more clearly, as they are line plots.

Let's see how to plot them now:

```
# Transform to data.frame type for base R plotting
txhouses_clean <- as.data.frame(txhouses_clean)
# Create a grid for plotting multiple histograms
par(mfrow = c(4,2))
# Plot histograms
for (var in colnames(txhouses_clean[2:8])) {
    hist(txhouses_clean[,var],
                col= "blue",
                main= paste("Histogram of", var),
                border="white" )
}
```

We will get the following results:

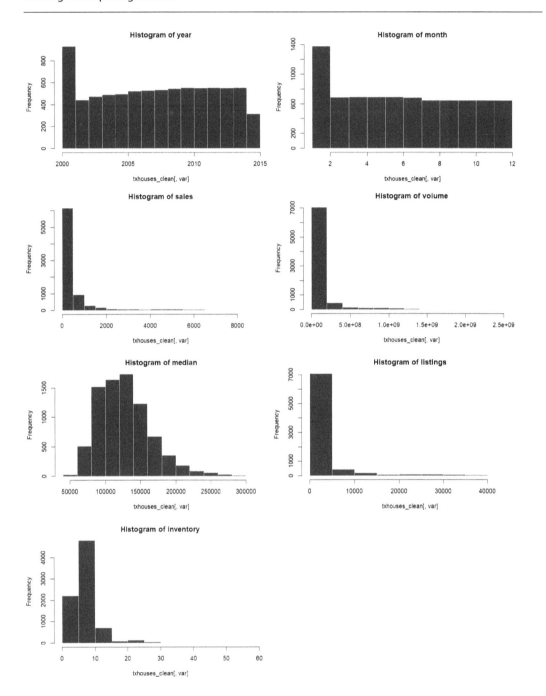

Figure 2.18 – Histograms for the variables in our dataset

The output of our code is shown in *Figure 2.18*. The insights it brings up are as follows:

- All the variables are right-skewed; that is, there is a high concentration of lower values, and the tail is longer to the right, indicating possible outliers.

- The `year` and `month` variables are constants; as we saw before, most of the cities have information available for every year and month.

- The `sales` variable is concentrated under 500, highly right-skewed with some outliers of up to 8,000 sales in a month.

- The `volume` variable is also right-skewed, with outliers.

- The `median` sales value is concentrated around $125k to $130k.

- The `listings` variable is concentrated around values under 5,000 houses available by observation.

- The `inventory` variable time is concentrated around 5 to 10 months.

After reading and understanding these histograms, we can conclude one important thing: the data has outliers. As we can see, there are concentrations but also long right tails for some variables, which indicates that a possible regression model would have much more material to capture patterns from on the left side of our distributions, making it harder for it to *learn and predict* from values that are too far away from those concentrations.

In regression models, we calculate the coefficients of a linear equation that explain the relationship between two or more numeric variables within a certain range. Now, look at *Figure 2.19*. Would it be safe to estimate the value of **P** using the linear regression created for the black points?

Maybe it would give us a fair prediction. But the fact is that it would be an extrapolation (`https://online.stat.psu.edu/stat501/lesson/12/12.8`) of the regression. This happens when you use a linear regression equation for points within a certain range to predict points that are not within that range. We don't know whether the red line will keep being a line up until **P**, or whether it will bend somewhere. **P** may be a participant of another distribution that would require another regression model. That is why we must take care of outliers and why it would not be the best decision to use that prediction model.

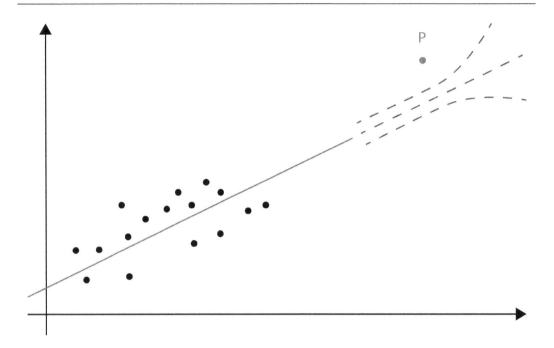

Figure 2.19 – Outlier and extrapolation

A good way to check outliers is using boxplots. Let's plot some using the following code:

```
# Create a grid for plotting multiple boxplots
par(mfrow = c(3,2))
# Plot Boxplots
for (var in colnames(txhouses_clean[4:8])) {
    boxplot(txhouses_clean[,var],
            col= "blue",
            main= paste("Boxplot of", var)    )
}
```

You will get the following output:

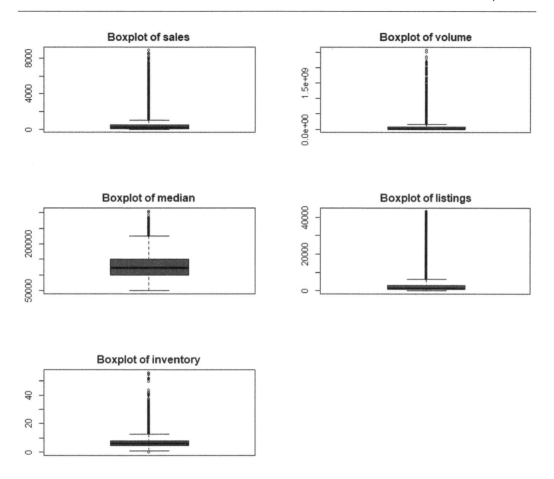

Figure 2.20 – Boxplots showing many outliers in the txhousing dataset

The boxplot brings a horizontal line at the top of the *whisker* to represent the maximum expected value of that variable. Any data point over that is considered an outlier, and we see a lot of them in our dataset.

Certainly, they require treatment, just like the missing values. If there are too many, we can separate those points and analyze them as a new dataset and even generate a different model for them. Likewise, we can use the imputation of upper and lower cap values or just remove them. The decision will vary from project to project and must be taken by the data scientist. In the *Further reading* section at the end of this chapter, there's a good source where you can study more methods of handling outliers. For the sake of space, I will skip them in this little project.

Visualizations

To end this EDA, we must plot some graphics. Visualizations are a must in any Data Science project because they bring to light patterns and insights that tables cannot show. The human brain is much more visual than auditory or textual. MIT neuroscientists have found that the brain can identify images seen for as little as 13 milliseconds (`https://news.mit.edu/2014/in-the-blink-of-an-eye-0116`). So, when we present graphics to our audience, they can connect to the information at a faster pace.

Our goal now is to check for possible relationships between variables that will help us to understand the data and possibly build a model in the future.

Keeping our main goal in mind, where the `median` sales price is our target, we can carry out exploration around that variable. As we have already plotted univariate graphics (histograms and boxplots), we can begin with a heatmap to check the correlations between the variables. After all, the R-squared of a regression model is nothing more than the square of the correlation:

```
# Create a correlation matrix excluding city
CM = cor(txhouses_no_outliers[, -c(2,9)])


#Plot the correlation heatmap
corrplot(CM, method = "square", type = "lower",
                  diag = FALSE, addCoef.col = "black")
```

The code publishes the following plot:

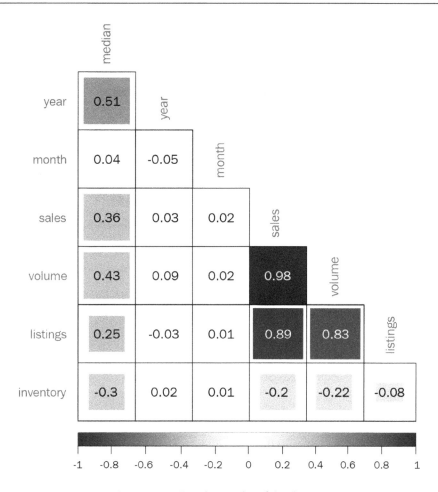

Figure 2.21 – Correlation plot of the dataset

As we can see in *Figure 2.21*, considering median as the target variable, the best correlations are with year and volume. But there is no variable that strongly explains the variance of median. We can see that volume, listings, and sales are multicollinear, so just one of them should stay in the model. The rest should be removed before modeling.

Another good graphic for EDA when we are looking at continuous variables is the **scatterplot**, one of the best options for visualizing the relationship between two variables. The pairs() function can do that for us in a single line of code:

```
#Plotting all the scatterplot for pairs of variables
vars_for_scatter <- c("median", "sales", "volume", "listings" ,
"inventory")
pairs(txhouses_no_outliers[,vars_for_scatter])
```

The code creates the following visualization:

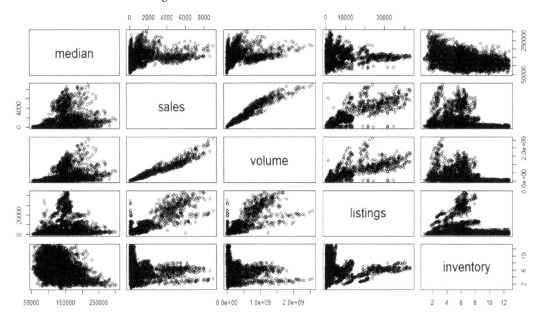

Figure 2.22 – Scatterplots generated in pairs

Figure 2.22 brings us interesting information. Look how the very linear plot of `sales` and `volume` confirms the high correlation. If you add both to your model, you are distorting it because the same variance is explained by two different variables. It's redundant information that could turn into noise. Notice how `median` does not show a clear linear relationship with any of those variables, which makes it more challenging to choose the best features for modeling.

With this pairs plot, we end this EDA. In conclusion, we could send our client, the real estate agency, the following brief:

- We have nine variables, one qualitative and eight quantitative.
- 17% of the observations have at least one missing piece information. The `listings` and `inventory` variables are the worst, with 10.6% and 11.2% of the data missing, respectively. Some cities have entire years with all NAs, which could affect the modeling for those locations.
- The dataset brings a lot of outliers, and future conversations must occur to define whether those can be discarded or whether there should be a model for them.
- If we want to predict the median sales price, the variables that hold a stronger correlation are `year`, `volume`, and `inventory`.

Before we close this chapter, the next section shows another option for acquiring data. The internet is a great place to collect data from good sources, such as many organizations and governments that provide open datasets for study.

So, for those interested in learning the basic steps to perform web scraping, keep reading. If you are not yet interested in this topic, you can skip this section without harm to the book's overall sequence.

Basic Web Scraping

Web scraping is another common way to acquire data. Given that we are in the information era, the quantity of data available online is enormous. Currently, around 2.5 quintillion bytes (2.5 followed by 18 zeros) is produced every day. Some experts estimate that 90% of all the data ever produced by humanity was generated within the last 2 years (`https://cloudtweaks.com/2015/03/how-much-data-is-produced-every-day/`). So, if you need data, there's plenty around.

But before I move on with this section, I'd like to include this disclaimer. Although web scraping is legal, there are some rules and ethical standards that we must follow when scraping data from a website.

First, we must respect private data and never use it without permission, as it belongs to other people or companies. The rule is to always look for public data, especially if it is provided via APIs, because that means that the datasets were reviewed and made available for use under defined rules.

Second, another good practice is to always check the `robots.txt` file of the page you are about to scrape data from. Most modern websites have them, and they can be accessed by writing `thewebsitename.com/robots.txt`. For our example, we are going to scrape data from Wikipedia, so we can look at `www.wikipedia.com/robots.txt` and make sure they allow that. Also, check the requests you have to follow in order not to overwhelm the servers and retain good service for everyone on the web.

Begin by loading the required libraries:

```
library(rvest)
library(tidyverse)
```

Now we will use `rvest` to start scraping and manipulating the data. Here is a screenshot of the page we are interested in from Wikipedia that presents the GDPs of different countries.

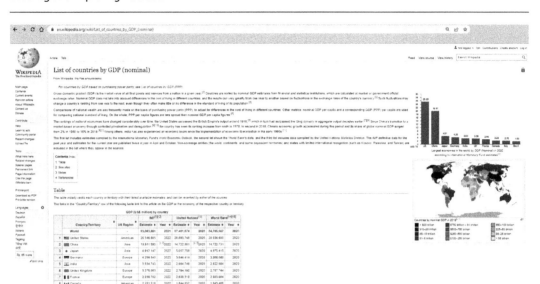

Figure 2.23 – List of countries by GDP

As seen in *Figure 2.23*, there is text and a table on this page (`https://en.wikipedia.org/wiki/List_of_countries_by_GDP_(nominal)`). Our main objective is to scrape the table. But we will go over other basic functions before we get there:

```
# Target Page
page <- "https://en.wikipedia.org/wiki/List_of_countries_by_
GDP_(nominal)"
# Read the page and store in a variable
gdp <- rvest::read_html(page)
```

In the preceding code snippet, we defined a URL to scrape and stored it in the gdp variable. The output will be an XML-type object with a list of two objects: the head and body of the page. We can use other functions to extract only what is interesting to us from that list, such as the first paragraph of the page, if you are interested in the text:

```
# Extract only paragraph 1 from the web page
p1 <- gdp %>% html_elements("p") %>% html_text()


# We are using [2] because the [1] is just a space break.
p1[2]
```

```
[1] "Gross domestic product (GDP) is the market value of all
final goods and services from a nation in a given year.[2]
Countries are sorted by nominal GDP estimates from financial
and statistical institutions, which are calculated at market or
government official exchange rates. Nominal GDP does not take
into account differences in the cost of living in different
countries, and the results can vary greatly from one year to
another based on fluctuations in the exchange rates of the
country's currency.[3] Such fluctuations may change a country's
ranking from one year to the next, even though they often
make little or no difference in the standard of living of its
population.[4]"
```

Now, let's move on to our goal: to get the table with the countries and GDPs, the real data.

To do that, we need to tell the rvest parser the exact point to look for in that table. And here, a basic knowledge of HTML will help you. Here is what you need to do:

1. Go to the page on a web browser. I recommend that you use Chrome.
2. Right-click on the table and select **Inspect**.
3. Another section will open in your browser, and you will see the HTML code.
4. Hover over the code to find the piece of code that highlights the table completely.
5. Right-click on that piece of code.
6. Select **Copy | Copy XPath**.

Refer to *Figure 2.24* while following the steps. Observe that the table is highlighted in blue when we hover over the code that refers to it.

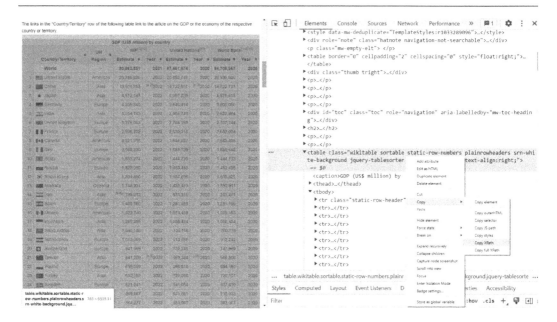

Figure 2.24 – Finding the XPath in HTML code

With the XPath collected from the HTML code, use the following code snippet to extract the page content:

```
# Extract the table from the page
gdp_df <- gdp %>%
    html_elements(xpath = "//*[@id='mw-content-text']/div[1]/
table[2]") %>%
    html_table() %>%
    .[[1]]
# View the table
View(gdp_df)
```

We have created another variable named `gdp_df`, pointed to the XPath found in the HTML code, and added the `html_table()` function to extract a table from a web page. Finally, we added `.[[1]]` to the pipe because it returns a list for us, but we are only interested in the `1` element, which is the table itself. `View(gdp_df)` allows you to use RStudio's viewer to look at the table.

And this is the resultant view.

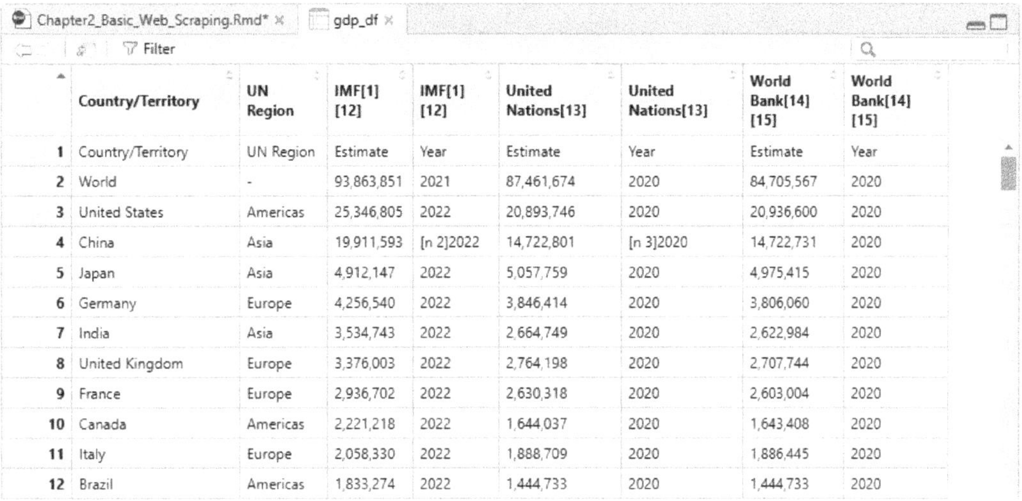

Chapter2_Basic_Web_Scraping.Rmd* × gdp_df ×

Filter

	Country/Territory	UN Region	IMF[1] [12]	IMF[1] [12]	United Nations[13]	United Nations[13]	World Bank[14] [15]	World Bank[14] [15]
1	Country/Territory	UN Region	Estimate	Year	Estimate	Year	Estimate	Year
2	World	-	93,863,851	2021	87,461,674	2020	84,705,567	2020
3	United States	Americas	25,346,805	2022	20,893,746	2020	20,936,600	2020
4	China	Asia	19,911,593	[n 2]2022	14,722,801	[n 3]2020	14,722,731	2020
5	Japan	Asia	4,912,147	2022	5,057,759	2020	4,975,415	2020
6	Germany	Europe	4,256,540	2022	3,846,414	2020	3,806,060	2020
7	India	Asia	3,534,743	2022	2,664,749	2020	2,622,984	2020
8	United Kingdom	Europe	3,376,003	2022	2,764,198	2020	2,707,744	2020
9	France	Europe	2,936,702	2022	2,630,318	2020	2,603,004	2020
10	Canada	Americas	2,221,218	2022	1,644,037	2020	1,643,408	2020
11	Italy	Europe	2,058,330	2022	1,888,709	2020	1,886,445	2020
12	Brazil	Americas	1,833,274	2022	1,444,733	2020	1,444,733	2020

Figure 2.25 – View of the table scraped from Wikipedia

Figure 2.25 shows the resulting dataset. With just a few lines of code, the data was loaded into RStudio and is ready for wrangling.

Web scraping is a useful resource, but it must be used with caution, following the best practices listed at the beginning of this section. Let's move on.

Getting data from an API

As advised previously, pulling data from an API is usually recommended over scraping. This is because the rules of using an API are easier to understand. Authorization codes are available that allow you to pull the data and certify that you are following best practices.

We are going to pull the MTS data from the US Treasury API, where the financial data is made public for everyone who needs it (`https://fiscaldata.treasury.gov/api-documentation/`).

An API works somewhat like a folder and file structure. The datasets are stored in **JavaScript Object Notation** (**JSON**) format and are accessible via a URL address. It is the job of the API developers to provide you with the keys to access their data. In general, there will be a base address, which is comparable to the *folder* in our analogy, and an endpoint, which can be thought of as similar to the *files*. For this API, the base URL is `https://api.fiscaldata.treasury.gov/services/api/fiscal_service` and the chosen endpoint is `/v1/accounting/mts/mts_table_1`. When you gather both, you have the URL to access the MTS report. Here is an extract of what you will find if you copy and paste the address into your browser:

```
{"data": [{"record_date": "2015-03-31","parent_id":
"null","classification_id": "12662418","classification_desc":
```

```
"FY 2014","current_month_gross_rcpt_amt": "null","current_
month_gross_outly_amt": "null","current_month_dfct_sur_amt":
"null","table_nbr": "1","src_line_nbr": "1","print_order_nbr":
"1","line_code_nbr": "10","data_type_cd": "S","record_type_
cd": "SL","sequence_level_nbr": "1","sequence_number_cd":
"1","record_fiscal_year": "2015","record_fiscal_quarter":
"2","record_calendar_year": "2015","record_calendar_quarter":
"1","record_calendar_month": "03","record_calendar_day": "31"},
```

It is not very pretty, but we will make it look better after parsing it. The code is as follows:

```
# Load libraries
library("httr")
library("jsonlite")
# Point to the URL wanted
url = "https://api.fiscaldata.treasury.gov/services/api/fiscal_
service/v1/accounting/mts/mts_table_1"
# Get the URL content
treasury_api <- GET(url)
```

The GET() function will extract the content from the web. Now, we have a list with 10 objects in the treasury_api variable. Our next step is to extract the content of the report and create a data frame out of it:

```
# Transforming the results to text
result <- content(treasury_api,"text", encoding = "UTF-8")
# Parsing data in JSON
df_json <- fromJSON(result, flatten = TRUE)
# Store as data frame
df <- as.data.frame(df_json$data)
```

The result variable gets the content and transforms it into JSON text that is very similar to the preceding text, using the UTF-8 accent encoding – one of the most used. That is performed with the content() function. Next, we parsed the JSON data and stored that in df_json. Finally, we got the report from that object and transformed it into a data frame, assigning it to the df variable.

You can see the result in *Figure 2.26*.

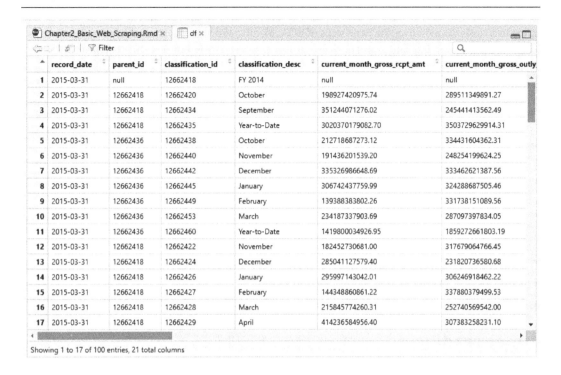

Figure 2.26 – Data frame pulled and parsed from an API

We are at the end of this chapter, and you can now load a dataset and perform a basic EDA for a Data Science project. Every Data Science project begins with data. That's the core part of our job. We use a mix of computer science, statistics, programming, math, and business knowledge to extract useful information from data, to support executive decisions.

Summary

We began *Chapter 2* by learning how to acquire data, using native R datasets or loading it from the popular CSV format, and how to customize the dataset even during the importing data phase, such as deciding on the number of rows to load. Then we briefly explained the difference between a data frame and Tibble format. They serve the same purpose and basically do the same things, but Tibbles bring some enhancements and are more suited to the modern world, and work much better with the tidyverse package in R.

Next, we advanced to more sophisticated ways to bring data to your R session by using web scraping or capturing datasets from a public API. As we live in a world where many businesses and salespeople work with Microsoft Excel, it is important to know how to save a file as a CSV. That was also covered in this chapter.

Coming to a close, we went over the basic steps of EDA: loading and viewing data, calculating descriptive statistics, handling missing values and outliers, and plotting data distributions and visualizations. We did all of that in the context of a working project for a real estate agency, making the example similar to a real-life Data Science project.

Data visualization is crucial for good data exploration. Thus, in the next chapter, we will learn more about data visualization with the built-in resources from R, enabling you to create the main kinds of plots, as well as customize their colors, titles, and labels.

Exercises

1. List three ways to load external data in RStudio.

2. How can I load just the first 2 columns and 10 rows of a dataset in R?

3. Revisit the list of differences between data frames and Tibbles.

4. How do you save a file in CSV format from RStudio?

5. What are the five main steps of a basic EDA?

Further reading

- Datasets:

    If you want to explore other sample datasets contained in the `utils` library, just write `data()` in the *Console* screen and it will display all the 104 datasets available. If you want to load it with a different variable name, use this command:

    ```
    # Load the dataset to a different variable name
    library(datasets)
    df <- datasets::Orange
    ```

- Tibbles:

    You can learn more about other nice features of Tibbles by entering `vignette("tibble")` in the R *Console* screen.

- HTML:

    Get familiar with HTML at this page: `https://www.w3schools.com/html/`.

- Outliers:

    `https://www.analyticsvidhya.com/blog/2021/05/detecting-and-treating-outliers-treating-the-odd-one-out/`

    `https://tinyurl.com/2p82vtb8`

    `https://statsandr.com/blog/outliers-detection-in-r/`

3
Basic Data Visualization

Perhaps I am too bold for saying this, but I don't think there is a point in wrangling or analyzing data if you are not going to visualize it. As previously mentioned, the human brain is so much better at understanding images than numbers or words. Furthermore, in the era of huge amounts of data, presenting tables would not be the most interesting way to identify or even just visualize patterns.

Data visualization means presenting information in a visual format, such as a graphic, a map, or another visual way of encoding data, such as an infographic, for instance, which is a combination of graphics, text, and other elements that help you to tell a story and transmit a message.

In this chapter, we are going to see basic data visualization using the native plotting capabilities of RStudio. We will start with plots of a single variable, which are the best way to visualize distributions. Included in this group are histograms, boxplots, and density plots.

Moving forward, we will go over two variate graphics that are important for checking relationships between two variables. Scatterplots, bar, and column graphics, as well as line plots, are part of this group.

Finally, we will close the chapter by plotting graphics with more than two variables. In this case, color, size, or the type of marker can become a way to differentiate data.

In this chapter, we will cover the following main topics:

- Data visualization
- Creating single-variable plots
- Creating two-variable plots
- Working with multiple variables

Technical requirements

We will use the dataset from Motor Trends Cars, widely known as mtcars. This is a good dataset for demonstration purposes.

All the code can be found in the book's GitHub repository: `https://github.com/ PacktPublishing/Data-Wrangling-with-R/tree/main/Part1/Chapter3`.

The libraries needed for the chapter are as follows:

```
library(datasets)
library(tidyverse)
# Dataset Load
data("mtcars")
```

Data visualization

Data visualization is part of every Data Science project, as it supports the ideas and concepts behind the theory. Graphics help us to tell the story in a better way than just numbers or text. This book is a good example if you think it through. How difficult would it be to explain what each kind of visual is if I could not show you a picture, right?

In this chapter, we will work with visualization using functions from what we call Base-R, a neat alias for the functions that come native in R. And before you wonder why we don't go straight to **ggplot2** – possibly the best and most well-known way to plot graphics in R – I would say that we're learning Base-R plotting because it does not require installation. Ergo, no dependencies, no conflicts, and no errors, making it your go-to option for a quick visual.

To be more specific about data visualization, let's go over a few good tips to build good graphics, according to Cole Nussbaumer Knaflic in *Storytelling with Data: A Data Visualization Guide for Business Professionals*:

- **Know your variables**: It is important to know whether your data is quantitative (numeric) or qualitative (categorical), so you know which graphic types should be considered.

- **Know your audience**: Knowing who will visualize the graphic can lead to better color, size, title, and subtitle choices – for example, using company colors for the graphics.

- **Choose the right representation**: Each kind of graphic will be better suited for different variable types. A line plot is better to show continuous data threads, such as time series. Bar plots are the best choice for category frequency.

- **Use other elements**: You can use other elements such as colors, size, text, title, and subtitle to drive the attention of your audience to specific points of the graphic.

- **Tell a story**: Using a combination of all the bullets just listed, you transform your graphic into a story, making it easier for the public to engage with the message.

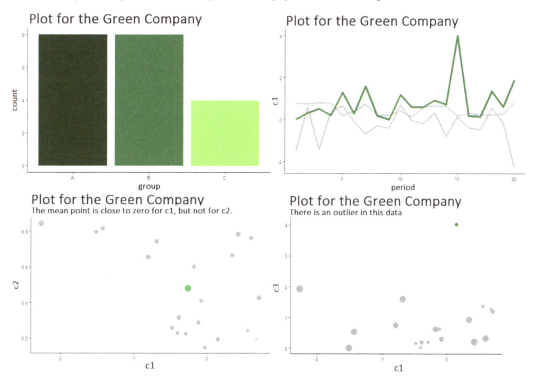

Figure 3.1 – Examples of graphics

Knowing your data, your audience, and the message you want to send with the plot will open a set of options to choose from. Graphics tell stories, and each variable is like a character that plays a role.

Next, we will learn the main kinds of plots for one variable.

Creating single-variable plots

Single-variable plots are mostly used to visualize the distribution of a variable. Using these kinds of graphics, it is possible to understand more of your dataset, evaluate where there is data concentration, data symmetry, or skewness, and visualize how the data behaves in comparison to the mean and detect patterns.

Dataset

The dataset chosen for this chapter comes from the `datasets` library; it is named *mtcars*. It is a widely known toy dataset for you to play with to learn coding and Data Science. For our goal here, which is understanding how to create each graphic, it presents itself as one of the best options because it is about a common subject (cars) and it has many variable numerical and categorical types for us to create different visualizations. If you want to know more about the variables, feel free to write `help("mtcars")` on your console in R. To load the dataset into an R session, just use the code that follows:

```
data("mtcars")
```

Histograms

The first univariate graphic to be presented is probably one of the most known and used, the histogram. A histogram organizes data in value ranges, also referred to as bins. For each bin, there is a count of how many data points fall into that range, making it easy to quickly identify the values around where the data is concentrated.

The dataset we are working with is about cars. The `mpg` variable, which should be read as miles per gallon, shows the fuel efficiency of each observation. We will look at its distribution to determine what the most common `mpg` value for the cars in this dataset is. Let's plot our first graphic, using the `hist()` function. The *x* axis is the miles per gallon variable, or just `mpg`:

```
# Histogram of miles per gallon
hist(mtcars$mpg)
```

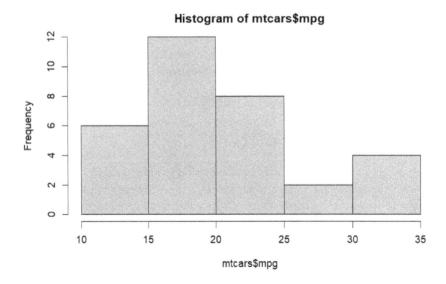

Figure 3.2 – A basic histogram

The most basic histogram that we can plot is this very simple single line of code. As you can see in *Figure 3.2*, it comes with all the default values, such as the *X* label equalling the `dataset$variable` name, the title being *Histogram of the variable name*, the number of bins according to the default calculation of the software, and the gray colored bars. It is pretty good compared to the basic plots we see in other programming languages.

Let's see how we can customize it a little, and after that, we can move on to other kinds of graphics:

```
#Custom histogram
hist(mtcars$mpg, col= "royalblue",
     main="Histogram of Miles per Gallon (MPG)", xlab = "MPG",
     breaks= "FD")
```

We used the same function, `hist()`, but we now add the `col="royalblue"` parameter for color, `main="Histogram of Miles per Gallon (MPG)"` for the title of the graphic, and the label for the *x* axis is set by `xlab= "MPG"`. The `breaks="FD"` parameter is the number of bins we will see for the histogram, for which we have chosen the *Freedman-Diaconis* calculation formula in this case (see `https://tinyurl.com/yw6e4s28`).

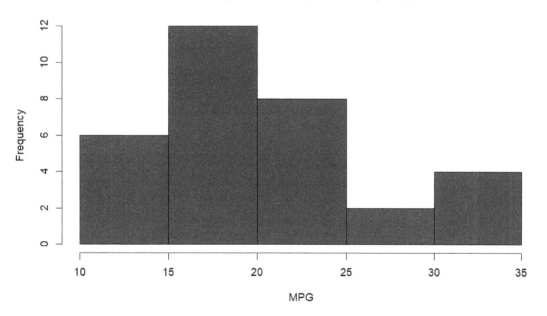

Figure 3.3 – A customized histogram

This looks much better. In the version in *Figure 3.3*, the graphic looks more professional.

As for the interpretation, it is just as previously explained. The bars represent the frequency count of observations that fall in each bin. For instance, the majority of cars have a **Miles per Gallon** (**MPG**) metric between 15 and 20 mpg – 12 times. Between 25 and 30, we see only 2 observations though. It is expected that we see fewer cars making more MPG, as fuel efficiency is a limitation.

Histograms are a great resource to visualize the distribution of a single variable, but boxplots can be too. If we want to see whether there is any outlier observation in this histogram, we can confirm that with a boxplot. Let's see how.

Boxplots

A boxplot is another way to visually represent a numerical variable distribution. The difference between a boxplot and a histogram is that the boxplot will show the variance using the quartiles.

Looking at *Figure 3.4*, an illustration of a boxplot, you can quickly identify the first quartile (**Q1**) where 25% of the data is below it, the median quartile (**Q2**) at 50%, and the third quartile (**Q3**) that captures 75% of the data. The boxplot also informs us of the calculated maximum and minimum values that are not considered an outlier (the lines on each side, or whiskers). The calculation is done using the **Inter Quartile Range** (**IQR**) – that is, the difference between **Q3** and **Q1**. Both formulas are embedded in the figure. And that leads us to this great feature of the boxplots – they plot the outliers from our data, something that no other plot will do as well. Any point past the whiskers will be marked as a possible anomaly, a value that is so much lower or higher than the bulk of the data.

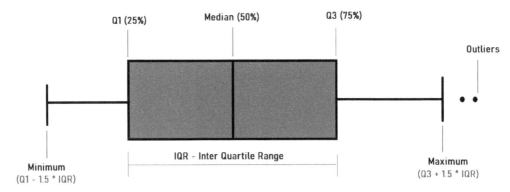

Figure 3.4 – Parts of a boxplot

Now that we know what a boxplot is, let's go back to our exploration question: does any car have a fuel efficiency that is much higher or lower than the others? We will discover that by plotting the next boxplot:

```
# Plot Boxplot of MPG
boxplot(mtcars$mpg, col="royalblue",
        main= "Boxplot of Miles per Gallon",
        ylab= "Miles per Gallon")
```

The function to plot boxplots is, intuitively, its name – `boxplot()`. Just like for the histograms, we are adding color with `col`, the title with `main`, and setting up the *y*-axis label with `ylab`. The onscreen result is shown here.

Boxplot of Miles per Gallon (MPG)

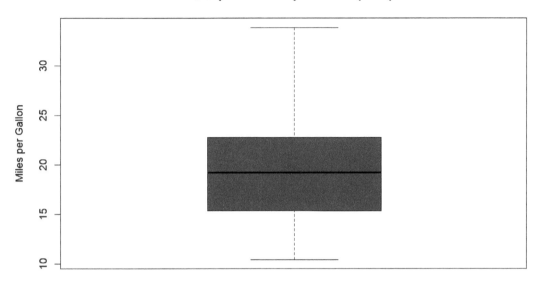

Figure 3.5 – The boxplot of MPG

That is a great visual in *Figure 3.5*. From all the 32 cars in the dataset, the median is around 19 and there is no outlier presented. The Q3 is 22.8, meaning that 75% of our observations are cars that can't make more than 23 MPG. I am sure you are starting to see how powerful the visualizations can be to extract insights from data.

Boxplots can be really useful when comparing two sets or groups of data within the same variable. For example, let's say during our **Exploratory Data Analysis** (**EDA**) we raise the question, *Do cars with automatic transmission make less miles per gallon than the manual transmission vehicles?* To answer that question, we can plot two boxplots side by side and compare both groups. Let's see that in action:

```
# boxplot AM vs MPG
boxplot(mtcars$mpg[mtcars$am == 0],
        mtcars$mpg[mtcars$am == 1],
        col= c("coral", "royalblue"),
        names= c("Automatic", "Manual"),
        ylab= "MPG",
        main= "Types of Transmission and the Fuel Efficiency"
        )
```

In this code, we used a vector with the same number and order of the groups as the input for the `col` and `names` parameters. Given that we have two groups, the first element of the vector will go to the **Automatic** group and the second will go to **Manual**.

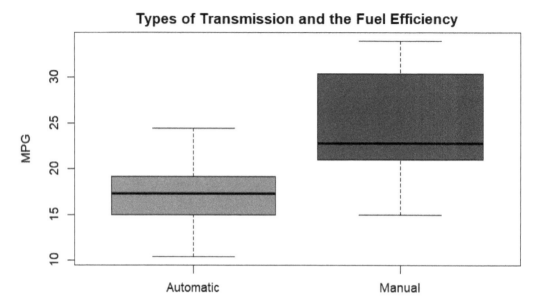

Figure 3.6 – A comparison between two groups using boxplots

It becomes clear after we plot the graphic, at least visually and before any statistical test proves it, that, on average, for this dataset, the cars with manual transmission provide a better fuel efficiency than the automatic ones.

There is still another way to plot boxplots side by side, using formula notation. R uses the tilde (~) to denote formula notation that usually means a comparison. You will see later in this book this usage when building models or other libraries that take advantage of the symbol. Base-R graphics are no different. So, let's use it for the next plot to discover the difference in fuel efficiency by the quantity of cylinders:

```
# Boxplot with formula notation
boxplot(mpg ~ cyl, data= mtcars,
        col= c("gold", "coral", "blue"),
        main= "Miles per Gallon by Number of Cylinders",
        xlab= "Cylinders", ylab= "Miles Per Gallon")
```

You can read mpg ~ cyl as how MPG responds to the changes in cylinders, with mpg being the response variable in this case. The conclusion is visible – the more cylinders in the engine, the lower its MPG.

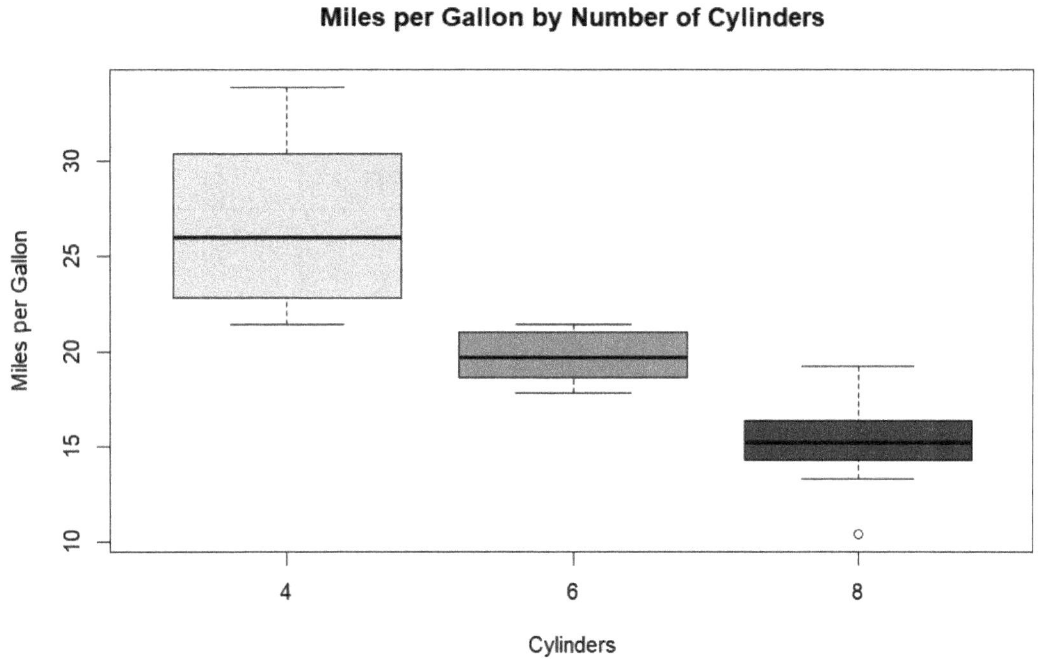
Miles per Gallon by Number of Cylinders
Miles per Gallon
10
15
20
25
30
4
6
8
Cylinders

Figure 3.7 – Boxplots created using formula notation

This is how we can use boxplots for quick checks of data distribution or comparisons between groups within our dataset. This graphic type provides so much information in a single plot, which is why statisticians usually like them.

Moving forward, we will learn about density plots as a good resource to look at distributions of variables.

Density plot

Now that we have looked at univariate plots, let's learn about the density plot. Just like histograms and boxplots, the density plot represents the distribution of a numeric variable. However, it does that showing the probability density function of the variable on the *y* axis. We will keep looking at MPG and its distribution, using another approach this time. The code is as follows:

```
# Density calculation
dprob <- density(mtcars$mpg)
# Plot
plot(dprob, main= "Density Plot of MPG")
# Fill shape
polygon(dprob, col="red")
```

To create a density plot with Base R, you first need to use the `density()` function to calculate the numbers to be plotted. Then, you can call `plot()`, which is the basic command to plot a line, with the new variable just created (`dprob`) as input. Finally, use the `polygon()` function to fill the shape of the curve.

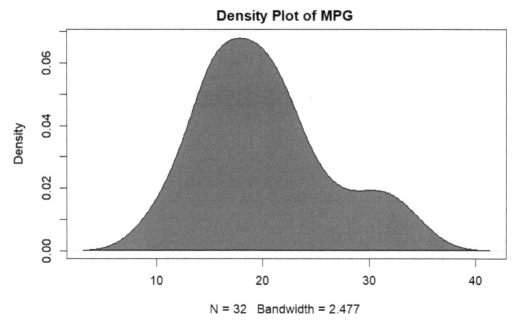

Figure 3.8 – A density plot

If you compare *Figure 3.8* with the histogram from *Figure 3.3*, you can see the similarities. The density plot is a smoothed version of the histogram. On the *y* axis, instead of seeing frequency, you now see the probability density, which is the probability of getting an *x* value between a range of *x* values. In other words, the probability of finding one observation with an `mpg` value of around 20 is about 6%.

Like the other types, we can also view comparisons plotted in the same figure, which are coded as follows:

```
# Density calculation
dp0 <- density(mtcars$mpg[mtcars$am == 0])
dp1 <- density(mtcars$mpg[mtcars$am == 1])
# Plot
plot(dp0, main= "Density Plot of MPG")
plot(dp1, main= "Density Plot of MPG", ylim= c(0,0.1) )
# Legend
legend(x=35, y=0.1 , c("Automatic", "Manual"), fill= c("red",
"blue"))
```

```
# Fill shape
polygon(dp0, col="red")
polygon(dp1, col="blue")
```

When plotting using Base R, most of the time to plot two lines in the same figure, the trick is to write the lines in sequence. RStudio will take care of putting them in the same plot. In the code for this plot, we used `density()` to create the calculations to be plotted, and then we called the `plot()` function twice, inputting the two density variables created, `dp0` and `dp1`. Other than that, the only different line is regarding the `legend()` function, where I added the `x` and `y` positions where the text will appear, along with a vector of the text for the legend, and the `fill` argument receives a vector for the colors of each curve.

The code will display the graph in *Figure 3.9*, where we can have a good comparison of the distribution of MPG for both types of transmission.

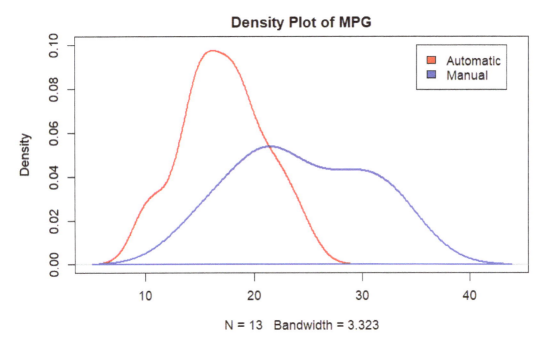

Figure 3.9 – Comparing density plots

If you know a little more about R, you might be wondering whether there is a way to create transparency to be able to completely see both plots when they are filled with color. The quick answer is that we will see that in *Chapter 10* about `ggplot2`, a much fancier data visualization library for R.

Meanwhile, it is now time to learn about plots with two variables, a valuable tool for comparisons and visualization of relationships.

Creating two-variable plots

We learned in elementary school that graphics are composed of an x axis and a y axis. Ergo, when something happens in X, there is a resultant change in Y. Understanding this simple phrase helps us to understand that a bi-variate graphic will represent the relationship between two variables of our dataset. If the x variable goes up, what happens to the y variable? What if x is constant?

Those and so many other questions can be raised and answered by visualizations of two variables. The first one we will see is the scatterplot. Let's move on.

Scatterplot

A scatterplot, also known as a points plot, is widely used for Data Science analysis, especially for regression analysis. Using this kind of plot, it is possible to see whether there is a linear relationship between x and y, for example, or to find another pattern in data. In the sequence of our exploration, we are interested in finding out the effect caused by the increase in horsepower on fuel efficiency. Does a more powerful car make more MPG or less? Let's discover this in the next scatterplot:

```
# Basic scatterplot
plot(x=mtcars$hp, y=mtcars$mpg,
     main= "Effect of Horsepower on Miles per Gallon (MPG)",
     xlab= "Horsepower", ylab= "MPG",
     col="royalblue", pch=16)
```

To plot the graphic, we can see that we used the plot () function once again, passing both x and y axes, a title using main, the xlab and ylab axes labels, the color using col, and the pch=16 parameter to determine the type of point we wanted for the visualization (filled circles).

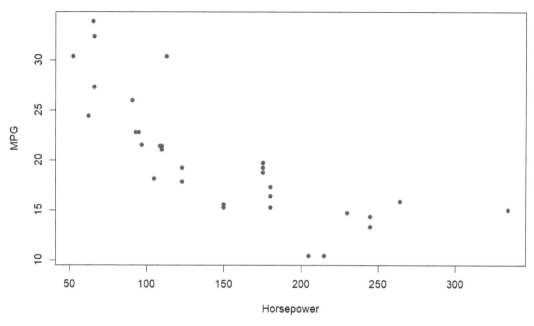

Figure 3.10 – A basic scatterplot of horsepower compared to MPG

The interpretation of *Figure 3.10* is pretty straightforward. For each point in *X* (**Horsepower**), there is an equivalent value in *Y* (**MPG**). We can see that, as the horsepower increases for the engines, MPG drops quickly. For cars in the 100 **Horsepower** (**HP**) range, MPG floats around 20. For the 250 HP cars, MPG doesn't go much over 15. Indeed, it makes a lot of sense. More potent cars are more likely to spend more fuel. It shows that our data is consistent with reality.

We can also change the size to 2 points using the `cex=2` parameter. See in *Figure 3.11* how the points are now bigger. You will also notice the same *negative* relationship between MPG and weight. When the weight increases, MPG decreases.

Note

When we have a relationship where *x* increases while *y* decreases or vice versa, we call it a negative relationship. If, on the flip side, both have the same behavior, increasing or decreasing together, that is a positive relationship.

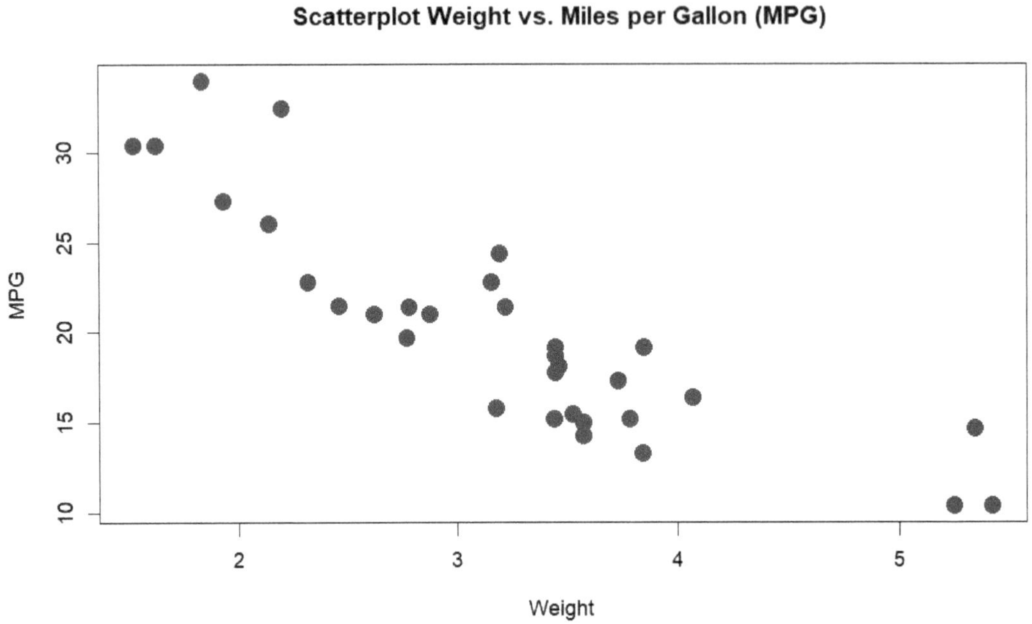

Scatterplot Weight vs. Miles per Gallon (MPG)

Figure 3.11 – Larger points for a scatterplot using the cex parameter

While scatterplots are a great fit to show the relationship between two numerical variables, it is not the best option for plotting categories. For that operation, bar plots come in handy for their simplicity and effectiveness.

Bar plot

A bar plot is a visual used to represent values related to categories. It can be the frequency, sum, mean, or any value related to a categorical variable. A bar plot and a column plot can be used interchangeably, although there is a difference in the layout of the graphic – that is, the column plot has categories on the x axis and the bar plot has categories on the y axis.

To create a bar plot, we should first determine which variables would fit better represented by a bar. Even though our 11 variables are all numerical, it does not mean that every one of them is quantitative. Numbers can represent quantities, orders, categories, or Boolean information, such as true or false.

Additionally, let's remind ourselves to always look at the documentation of the dataset by writing help("mtcars") on R's console, as it has the data dictionary to be referenced. We will find out that there are binary variables such as am, representing categories – 0 means automatic and 1 means manual transmission. Likewise, vs is the engine (0 = V-shaped and 1 = straight) or the number of cylinders (cyl).

Despite the fact that they are numbers, all those variables in fact represent categories or qualities of a car. We already know that the quantity of cylinders makes a difference in the fuel efficiency of cars. So now, we want to know how many cars there are in the dataset by the number of cylinders. A column plot can answer that question:

```
# Create a table count for each cylinder type
cyl_counts <- table(mtcars$cyl)
# Bar plot
barplot(cyl_counts, col="royalblue",
        main= "Number of Cars by Cylinder",
        xlab = "Cylinders", ylab = "Number of Cars")
```

The code starts with the creation of a `cyl_counts` variable that uses the `table()` function to count how many observations we have by cylinder (`table(mtcars$cyl)`). That is done first because the `barplot()` function requires a matrix or a vector as input. In simpler words, we must do the count first and then present the categories and values associated with the function. To create the graphic, we use the `barplot()` function with the `cyl_counts` variable just created, along with parameters for color and title that we are familiar with from previous code.

The result is displayed as follows.

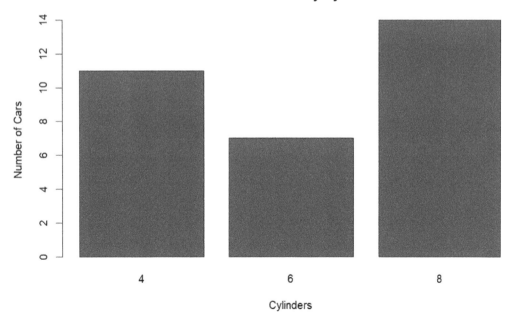

Figure 3.12 – A basic column plot

Figure 3.12 shows the basic column plot created with Base R functions. The result connects with what we have explored so far. Think about it – if more cylinders means less MPG, and we have more cars with a higher number of cylinders, then the average of mpg is expected to be lower. Therefore, it makes sense to see a skewed mpg histogram with more observations on the left side, with lower mpg values.

To change the orientation of the column to horizontal and plot a bar plot, we just need to add the `horiz=TRUE` parameter. To illustrate that kind of graphic, let us count the number of observations with straight line-shaped engines and V-shaped engines. The code is as follows:

```
# Create a table count for each engine type
v_eng <- table(mtcars$vs)
# Bar Plot
barplot(v_eng, main="Cars with V-Engine vs Straight",
        horiz=TRUE, col="royalblue",
        names= c("V-Shaped", "Straight"))
```

Observe that we did the same here – created a new variable to hold the counts by the engine and named it v_eng. Then, we passed this variable to the `barplot()` function plus title and color. The different parameters presented in the preceding code snippet are `horiz=TRUE`, to make the bars plotted horizontally aligned, and `names`, which receives a vector of the same length as the number of bars to display the labels on the *y* axis – two bars and two names. You can see the result in *Figure 3.13*.

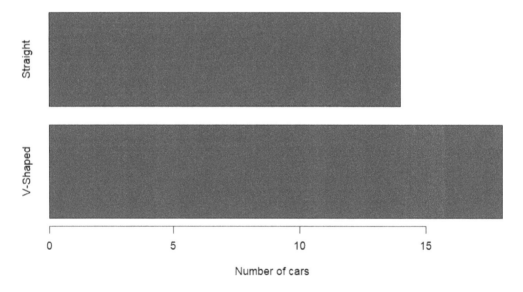

Figure 3.13 – A basic bar plot

Just a final note on bar plots – don't confuse them with histograms. They are not the same thing. While bar plots have a category on one axis and bring a **value** associated with it (that value can be a frequency count too), histograms bring the frequency by a **range of numeric values**.

Although bar plots can be used to show evolution through time, there is another type of graphic to consider as an option for that purpose – line plots.

Line plot

Line plots are a perfect fit to represent the evolution of one or more variables over time. If we want to plot trends or just want to compare how sales of store A are better or worse than that of store B for the last 5 years, a line plot is our graphic.

Analyzing our toy dataset, mtcars, we can see that its variables mostly call for other kinds of graphics. There is no time series data, for example, that would require a line plot. So, let's use another dataset just for the purpose of showing you a good line plot.

The dataset chosen is AirPassengers, a time series that shows the evolution of the number of people traveling by airplane from 1949 to 1960. For this visualization, all we want to look at is the trend line between those years registered in the dataset.

The code in the sequence is where we load the dataset and use the plot() function, providing it with the dataset, color, title, and the lwd=3 parameter, which configures the width of the line:

```
# Load dataset
data("AirPassengers")
# Plot line
plot(AirPassengers, col= "skyblue", lwd=3,
     main= "Evolution of the Number of Air Passengers from
1949-1960")
```

The result can be seen in *Figure 3.14.*

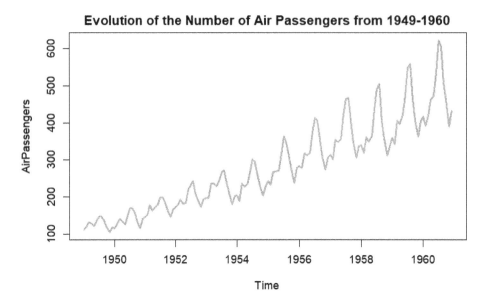

Figure 3.14 – A time series plotted as a line

As the dataset is a **time series** (**ts**) object type, we just use a plot and pass the name of the dataset, and RStudio takes care of the rest.

If you want to plot multiple lines on the same figure using Base R capabilities, the solution is to use the lines() function to add the extra series. For demonstration purposes, we can create three random variables to plot.

The code creates an x variable with a sequence from 1 to 12 and another three variables, y, y2, and y3, with random numbers from a normal distribution that can be generated with rnorm(). After that, we create a first layer plot with the plot() function, passing the *x* and *y* axes, the graphic type with type, the limits of the *y* axis with ylim, and the color. Next, we start adding the other two lines using the lines() function. You can control the line type using the type argument (where l means line, p means points, and b means both), the width with the lwd argument, and the line type, such as dashed or continuous, using lty:

```
# Random X and Y
x <- 1:12
y <- rnorm(12)+2
y2 <- rnorm(12)+3
y3 <- rnorm(12)+4
# Line Plot
plot(x,y, col="royalblue", type= "l", lwd=3, ylim=c(0,6))
```

```
# Add second line
lines(x,y2, col="coral", type= "b", lwd=3)
# Add third line
lines(x,y3, col="darkgreen", type= "l", lwd=3, lty = 2)
#Add legend
legend(x=11, y=6, c("y", "y2", "y3"),
fill=c("royalblue","coral", "darkgreen"))
```

The graphic returned after running this code is shown in *Figure 3.15*.

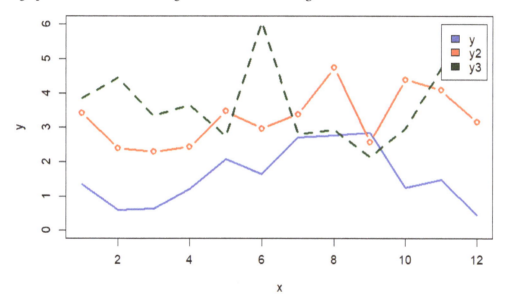

Figure 3.15 – Multiple lines plotted on the same figure

Certainly, *Figure 3.15* is busy enough, but the goal, in this case, is to show the possibility of multiple series plotting on the same figure.

Graphics are commonly plotted in two dimensions, but that does not mean that they bring only one variable for the *x* axis and one for the *y* axis. We can include other variables such as size and color in a graphic, making it a multivariate plot.

Working with multiple variables

A graphic can have more than two variables, not just what is plotted on the *x* and *y* axes. We can use colors, marker shapes, or sizes to differentiate data points and create a more complex visual. Look at these basic examples.

Scatterplots are the best fit for multiple variate plots, as the points can be changed to other shapes, sizes, or colors and produce a very rich visual. Knowing that the number of cylinders (cyl) and horsepower (hp) affect directly the fuel efficiency of a car (mpg), a good exploration point is visualizing the effect of increasing cylinders and HP and observing how the fuel efficiency will respond. To perform the task, we plot a scatterplot that shows the relationship between the engine's HP with the MPG presented by the car. Then, we add the cylinder information as a third variable to control the size of the bubbles, making them larger or smaller, thus bringing more information to this graphic:

```
# Scatterplot 3 variables
plot(mtcars$hp, mtcars$mpg,
     col=mtcars$cyl, cex=mtcars$cyl/3,
     xlab="HP", ylab="MPG",
     main= "Effect of Horsepower and Number of Cylinders on the
Fuel Efficiency")
#Add legend
legend(260, 32, sort(unique(mtcars$cyl)),
       col=sort(unique(mtcars$cyl)),
       cex = sort(unique(mtcars$cyl)/6),
       pch=1, horiz = T, yjust = 0.5)
```

That is a lot of code, for sure, but we can break it up. We started by using the plot() function with the *x* and *y* values listed, and then we added color by cylinder when we wrote col= mtcars$cyl. The size is configured by the cex parameter, which also received the cylinder variable. Then, labels for *x* and *y* were added using xlab and ylab, and a title was added using main.

The legend() function creates the legend at the top right of the graphic. It takes in the *x* and *y* positions of the box and then the text itself, which, in this case, is a combination of the sort() and unique() functions, to create a sorted vector of the distinct cylinder sizes (4, 6, and 8). The same information was used for colors, so we will see matching names and colors. For the size of the circles in the legend, the cylinder values will be too big for the plot, so we divided them all by 6 (cex = sort(unique(mtcars$cyl)/6)). The pch=1 argument is for circles not filled, horiz=T is to create the legend box on the horizontal alignment, and yjust=0.5 is to centralize the values within the legend box.

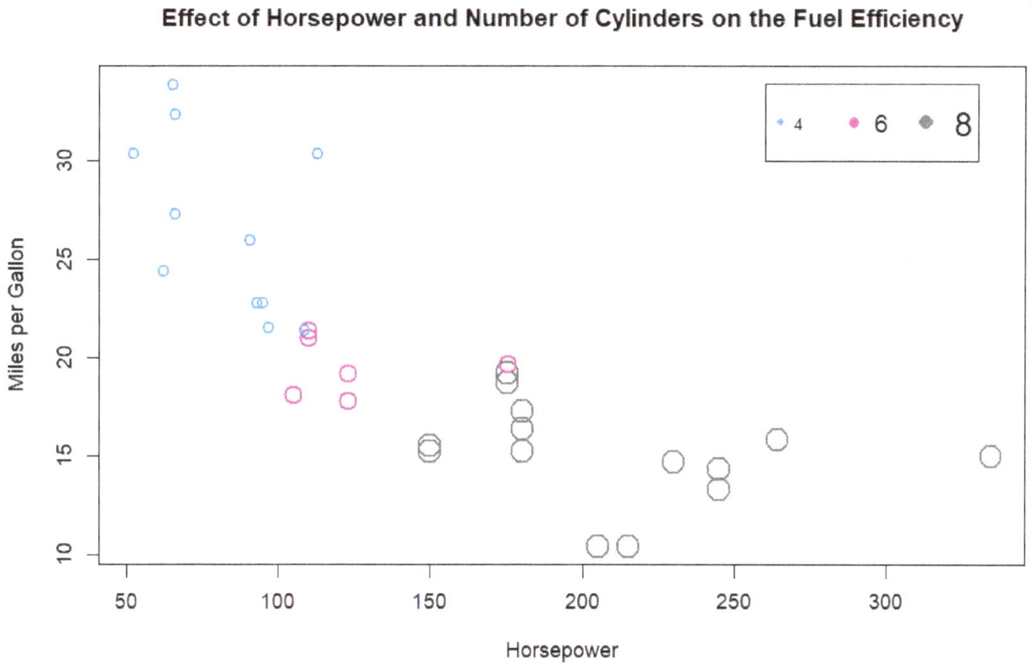

Effect of Horsepower and Number of Cylinders on the Fuel Efficiency

Figure 3.16 – Three variables plotted – cylinders differentiated by color

The result of that chunk of code we have just explained is in *Figure 3.16*. Adding an extra variable to the plot adds information to the figure, which requires more attention and focus from our audience. Just to make it clear, the small blue circles are cars with four cylinders, the medium-sized red circles are cars with six cylinders, and the bigger circles in gray are eight-cylinder cars. Now, we can answer our exploratory question by seeing that a higher HP and number of cylinders will burn more fuel, consequently lowering the MPG per car. To reach such a conclusion, we just look at the larger bubbles (eight-cylinder cars) floating closer to the bottom of the scatterplot, meaning a lower MPG.

When using multiple variate plots, we should be careful when using too much information to create a graphic. Look at *Figure 3.17*, a messy column plot combined with line plots, a caption, and two legends, which bombards us with information, making it more difficult to understand what message is being conveyed by the graphic.

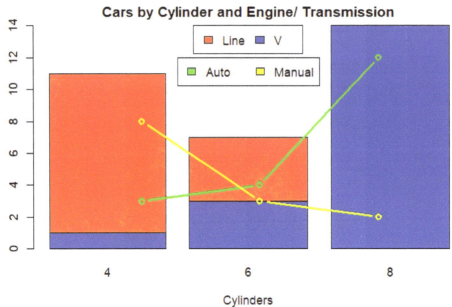

Figure 3.17 – An example of a polluted visual with too much information

Another thing to note is that we did not use the transparency feature (commonly known as *alpha* in ggplot2) because Base R plots don't have that as an add-on. We will see that parameter in use later in this book, with more capable libraries.

Plots side by side

Creating side-by-side plots is an interesting feature to compare scenarios or complement information. A good example of what works well in a side-by-side plot is a pre-post comparison – how it was versus how it is now.

Another nice use case is when you show information and want to complement it with another graphic, such as what we are about to do. Let's look at displacement – that is, the volume of fuel that cylinders can take at once – and analyze the relationship that has with the HP of a car. Do higher displacement numbers mean the car receives more power? Let's check:

```
# Setup grid with 1 row and 2 columns
par(mfrow= c(1,2))
# Plot graphics
# Plot (1,1)
plot(x= mtcars$disp, y= mtcars$hp,
```

```
        col="royalblue", pch=16,
        main="HP by Displacement")

# Plot (1,2)
hist(mtcars$disp, col="skyblue",
        main= "Histogram of Displacement")

# Reset plot grid to the original 1 by 1
par(mfrow= c(1,1))
# View if grid is reset. Should see 1 1
par("mfrow")
```

The previous code snippet features the par() function to set up a grid of spaces to receive graphics. In our case, we need one row and two columns, so we input a vector to the mfrow=c(1,2) parameter. Next, we will create the first plot using the plot() function with x, y, the color, the filled circle point type, and the title, in this order. The second plot comes in sequence. It is a histogram created with hist(), which receives the variable displacement, along with color and title parameters. To finish the snippet, we reset R's grid to the default one plot per row by calling par(mfrow= c(1,1)).

The result is shown as follows.

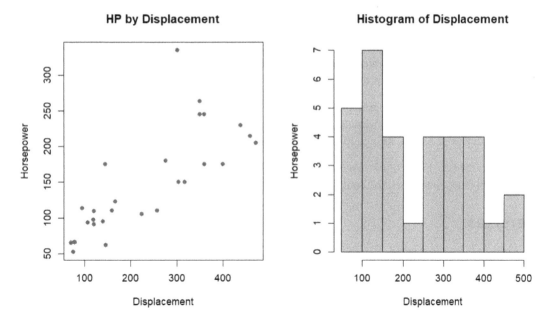

Figure 3.18 – Side-by-side graphics

In *Figure 3.18*, the graphic on the left-hand side shows how the increase of displacement (the volume that the cylinders can take of fuel) can influence the total HP of the car. As the displacement increases, HP follows along. It is a positive correlation. On the right-hand side, there is a plot of the histogram for displacement, informing us that the observations are slightly concentrated between 100 to 150, which confirms the same information on the scatterplot. There are some more points on the lower-left side, where the x axis for displacement is around 100 to 150. The graphics complement each other's information.

Plots side by side are a great use of space. They make quick comparisons easier, and they can be used during a presentation to support your message.

This chapter closes the first part of this book. We have been introduced to data-wrangling concepts, frameworks, exploration, and some visualization.

Summary

In this chapter, we learned what data visualization is and understood its importance for any project. Our brain is designed to quickly capture images; thus, graphics are more likely to be absorbed by an audience than by a table or text.

We introduced the basic types of single-variable plots – that is, histograms, which are commonly used to view the distribution of variables, and boxplots, which are especially good at detecting outliers.

In the sequence, we learned about graphics with two variables, such as scatterplots, that can show us how *x and y* are related and whether that relationship is positive or negative. We also learned about bar plots, a good representation of categorical variables because they are one of the simplest types of visual and easily understandable. Finally, we looked at line plots and how they are a great fit for continuous data and time series plots.

The chapter concluded with some examples of plots with many variables, with the scatterplot determined as the best option to create those visuals. All of that was created using nothing more than the native graphics library from Base R, known for its easy coding and eliminating the need for installations.

In the next part of the book, our journey continues with a deeper dive into the data wrangling topic, where we will study data types and transformations that can be made on them.

Exercises

1. Why do you need to know your variable types before plotting a graphic?

2. What is the best option to plot a graphic with a categorical variable?

3. If you want to detect outliers, which graphic can you plot?

4. Which kind of graphic can you use to plot a time series?

5. Why is the scatterplot a good option for multivariate plots?

Further reading

- Probability density:

  Stack Exchange explanation: `https://tinyurl.com/4ed7jp2k`

  Towards Data Science: `https://tinyurl.com/yfznae8x`

- Types of points for plots:

  Write `help("points")` on your RStudio console.

- Bar plot:

  `https://www.rdocumentation.org/packages/graphics/versions/3.6.2/topics/barplot`

  `https://www.statmethods.net/graphs/bar.html`

- Line plot:

  `https://www.rdocumentation.org/packages/graphics/versions/3.6.2/topics/plot`

- R Graphic Gallery:

  On this web page you can find code snippets for many graphics built with R:

  `https://r-graph-gallery.com/`

- *Storytelling with Data*, by Cole Nussbaumer Knaflic:

  `https://www.storytellingwithdata.com/`

- Code for this chapter in GitHub: `https://tinyurl.com/8mydmk95`

Part 2: Data Wrangling

This part includes the following chapters:

4
Working with Strings

Strings, in the programming world, are textual information: a single letter, a word, a phrase, or, more generally, anything that comes in between single or double quotes will be understood as a string by the computer once it is assigned to a variable. See the following code and comments:

```
# If not assigned to a variable, a text is just a comment.
"This is a text."


# These are strings
my_string1 <- "a"
my_string2 <- "Hello, World! I am learning!"
my_string1 <- "42"
```

The manipulation of strings is a good skill to have due to the amount of good data that is found on the internet in textual format. **Natural Language Processing (NLP)** is one of the largest areas in data science, and a lot of it relies on wrangling strings.

Most of what can be done with strings involves tasks such as the following:

- **Parsing**: Separating parts of the text that are divided by a pattern, extracting parts of it and combining words

- **Data mining**: Splitting the text into units (tokenization), finding the root or main meaning of a word with lemmatization, and stemming to find patterns and get a good insight into texts

In this chapter, we are going to focus on parsing, learning about the modern concepts for transforming textual variables. By the end of this chapter, you should be able to extract parts of strings, parse, and blend text into data frames. We will also work with regular expressions (*regexps*), which allow you to find textual patterns and highly specific strings with the power of customization.

We will cover the tools to use to find the frequency of a given word and build data summaries from textual data as well.

We will cover the following main topics:

- Introduction to `stringr`
- Working with regexps
- Creating frequency data summaries in R
- Text mining
- Factors

Introduction to stringr

There is a lot to learn about strings. Even though R is a language that was created with statistics in mind, it has developed a lot over the years and many libraries have emerged. As already mentioned, working with strings is a good skill to have given that you will often need to deal with these objects in your daily work as a data scientist. Sentiment analysis of clients or social media, comments analysis in feedback forms, the analysis of textual information scraped from the internet, or simply parsing a city name out of an address are some of the many tasks that can be part of a data wrangling request.

To code along with this chapter, make sure that you have installed and loaded the following libraries. Of them, you may be missing the Gutenberg package. Therefore, I suggest that you use `install.package("gutenbergr")` before trying to load it. We will use it for an exercise at the end of this chapter:

```
# Use install.packages("library_name") if you don't have it
installed
library(tidyverse)
library(stringr)
library(gutenbergr)
```

Let's start simple. To create a string, simply use single (') or double (") quotes. Most programming languages don't differentiate between them. R is no different, so feel free to pick whichever you like:

```
string1 <- "I am a string."
string2 <- "Me too."
string3 <- 'quote inside "quote", use single quote for the
string and double for the text.'
```

If you want to print a string, use `print()` or `writeLines()`. The difference is that `writeLines` won't bring any unwanted characters, such as quotation marks or escape characters such as the backslash (\). Notice that if we use `print` for `string1`, it will show double quotes. If we use `writeLines`, it won't:

```
# Printing
print(string1)
writeLines(string2)
writeLines(string3)
```

```
> # Printing
> print(string1)
[1] "I am a string"
> writeLines(string2)
me too
> writeLines(string3)
quote inside "quote", use single quotes for the string and
double for the text.
```

Now that we know how to create a string, let's work with the main functions from the `stringr` library. There are too many functions to go over all of them, but there is a link to the official documentation in the *Further reading* section if you want to explore more. In general, the syntax of `stringr` starts with `str_`, followed by the function name.

Detecting patterns

Detecting patterns is a common task when wrangling strings. You may want to know whether a phrase contains a certain word or whether a word contains a given letter.

The `str_detect()` function helps you detect patterns within a word or a phrase:

```
# Create a string
text <- "hello"
# Detect if the string has the letters "rt"
str_detect(text, "rt")
[1] FALSE
# Detect if the string has the word "world"
str_detect(text, "ll")
[1] TRUE
```

```
# Create a phrase
text <- "hello, world!"
# Detect if the string has the letters "ll"
str_detect(text, "world")
[1] TRUE
```

When we want to know whether a string starts with a specific pattern, we can use the `str_starts()` function. It can be useful if we have a data frame with an `"ID"` variable that uses a pattern such as `"ORDER-1234"` for purchase transactions and `"MAINT-1234"` for maintenance transactions. In that case, we could use the `str_starts()` function to extract the first part of the pattern. It will tell us which lines will be marked as purchases and which ones will be maintenance transactions:

```
# Create strings
my_id <- "ORDER-1234"
# Starts with
str_starts(string=my_id, pattern="ORDER")
[1] TRUE
```

When we have a vector of strings and we need to know the index of a pattern, we can use `str_which()`. The code is as follows:

```
# Create vector
shop_list <- c("fruit", "vegetable", "pasta")

# Index of "pasta"
str_which(string=shop_list, pattern="pasta")
[1] 3
```

We found `"pasta"` as the third element.

Similarly, the `str_locate()` function shows the start and end positions of patterns. In the following code, we will use the same vector. Note that the first two results are NA, as they do not match the *pasta* pattern, and the third result says that the word starts at the position of 1 and ends at the position of 5:

```
shop_list <- c("fruit", "vegetable", "pasta")
str_locate(shop_list, "pasta")
     start end
[1,]    NA  NA
[2,]    NA  NA
[3,]     1   5
```

As a good exercise to consolidate this concept, try changing `pasta` to `more pasta` in your code and see what happens.

To count how many times a pattern occurs in a text, use the `str_count()` function:

```
# Text
text <- "I want want want to count the repetitions of want in
this phrase. Do you want the same?"
# Count occurrences of want
str_count(text, "want")
[1] 5
```

Great. We found all five repetitions of the *want* pattern. Detecting, locating, and counting string patterns are functions to check whether a textual pattern exists within a text and to locate it in a phrase. Additionally, there are other functions to subset strings when you need to parse parts out of a text. Let's study them.

Subset strings

To subset strings means to extract a part of them for any purpose. If we know where to find the pattern in a text, it's easy. Revisiting our example about the ID with `"ORDER-1234"`, we know that the first five letters are the pattern we need, so we can just extract the `"ORDER"` or `"MAINT"` transaction type using the `str_sub()` function to return a substring, as the documentation refers to it. See here:

```
# Create strings
my_id <- "ORDER-1234"
# Subset characters from 1 to 5
str_sub(my_id, start=1, end=5)
[1] "ORDER"
```

The previous code returned only the first five characters from the string, as requested. However, suppose there is a list of transactions from which we only want to extract the ORDER entries and exclude the MAINT entries. The function to be used is `str_subset()`, which returns strings that contain a given pattern. To use it within `stringr`, see the following:

```
# Text
my_ids <- c("ORDER-1234", "ORDER-2234", "MAINT-1234",
"MAINT-2234")
# Return orders
str_subset(my_ids, "ORDER")
[1] "ORDER-1234" "ORDER-2234"
```

Notice that the result returns entire elements, not just parts of them.

Parsing text is a challenging task, and even more so if the text contains extra spaces. That is when functions for managing length become useful.

Managing lengths

Find out the size of your string using the `length()` function. Be aware that it considers spaces and punctuation:

```
# text
text <- "What is the size of this string?"
# Length
str_length(text)
[1] 32
```

We will present a couple of useful functions to trim and squash strings next. While `str_trim()` will remove the extra spaces at the end of a text but doesn't care about what is in the middle, `str_squish()` does both of these things:

```
# text
text <- " Text    to  be trimmed. "
#trim
str_trim(text, side="both")
[1] "Text    to  be trimmed."
```

```
# squish
str_squish(text)
[1] "Text to be trimmed."
```

Observe that the first function, for trimming, just removed the extra spaces at the ends of the text. `str_squish()` removed all the extra spaces.

Textual patterns won't match if they are not exactly the same, hence the importance of trimming and squishing. The same applies to uppercase and lowercase letters, as we are about to see.

Mutating strings

Another very common wrangling task with strings is changing the letter case. You can make quick conversions into **UPPERCASE**, **lowercase**, and **Title Case**. The function names are the names of the action to be performed, so they're very intuitive – `str_to_upper()`, `str_to_lower()`, and `str_to_title()`:

```
# text
text <- "Hello world."
# to UPPERCASE
str_to_upper(text)
[1] "HELLO WORLD."

# to lowercase
str_to_lower(text)
[1] "hello world."

# to Title Case
str_to_title(text)
[1] "Hello World."
```

These are simple but very useful functions.

Another important mutation task is replacing a pattern within a string. Notice that `str_replace_all()` will replace all the matching patterns at once, but `str_replace()` only replaces the first one found:

```
# text
text <- "Hello world. The world is beautiful!"

# Replace a pattern
str_replace(text, "world", "day")
[1] "Hello day. The world is beautiful!"

# Replace all the patterns at once
str_replace_all(text, "world", "day")
[1] "Hello day. The day is beautiful!"
```

Having a function to replace patterns is crucial. If you think for a second about how many times we repeat certain words in a text, it becomes logical that if we want to replace one of those words, we should be able to do so in one go.

When the task is to join words to create a text or a pattern, or split them to count words, for example, `stringr` is equipped for that too.

Joining and splitting

When working with text, it is common to split it or gather parts to form a new pattern. This can be done with the following functions. Let's join two words with `str_c()`:

```
# text
s1 <- "Hello"
s2 <- "world!"


# concatenate
str_c(s1, s2, sep=" ")
[1] "Hello world!"
```

If we can join strings, we can split them too. In this example, we used a space for the separator pattern, but we could use anything, such as a dash, slash, point, comma, and so on. A side note here: there are other concatenation functions in R. `c()` and `paste()` are both used to the same end – gathering elements – however, `str_c()` was constructed for this purpose and is equipped with vectorization for better performance and recycling, with the ability to concatenate vectors of different lengths:

```
#text
text <- "I am learning how to split strings"


# split
str_split(text, pattern=" ")
[[1]]
[1] "I"          "am"        "learning"
"how"        "to"        "split"       "strings"
```

Ordering strings

At times, it is interesting to order a vector of strings. Here is how to do so with `str_sort()`. To change the order, set the `decreasing` parameter to `TRUE` or `FALSE`:

```
# text
shop_list <- c("bananas", "strawberries", "avocado", "pasta")
```

```
# ordinate
str_sort(shop_list, decreasing = FALSE)
[1] "avocado"        "bananas"        "pasta"          "strawberries"
```

We have covered a lot so far. This library is so powerful – it enables you to wrangle text in many ways using the functions presented here. However, working with strings can become more challenging as a pattern becomes more specific. In these cases, we can add the power of regular expressions to `stringr`, enabling it to find highly specific textual patterns.

Working with regular expressions

Regexps are textual expressions – or codes – that empower programmers to create highly customized parsers for strings. Pretend that your client has a document with thousands of customer comments in it and they only want to extract the serial numbers of defective products. If you have ever seen a product serial number, you will know that it is a unique identifier of a product, generally composed of letters and numbers – for example, `XYW-001AB`. Well, using regexps, this task becomes easy, as it is possible to create a pattern that will fit that combination of letters and numbers for your client's products exactly, making parsing a quick task.

At first, when you look at a *regexp*, it will look like a weird code or something hard to learn, but as you begin practicing and using it more, the patterns naturally start to look easier to read and write. We will go step by step from zero to build our knowledge about regexps.

Learning the basics

The functions from `stringr` to use to view the matches for our *regexp* are `str_view()` – to return the first match – and `str_view_all()` – to return all the matches in the text. It is also important to remember that regexps are textual codes used to find textual patterns, so since we are using text to find text, there could be some conflict. The dot (`"."`) is a good example of those conflicts. It is used to denote any character when used in a *regexp*.

So, how do we look for a dot? In that situation, we can use an **escape character** – double backslashes (`\\`) – to tell the computer: *Hey, if I use \\., I am actually looking for a dot*. You can run this code – `?"'"` – to see other escape patterns:

```
# text
txt <- "Looking for a ."

# Regexp escape
str_view(txt, pattern="\\.")
Looking for a .
```

This is how you will see the result on your screen – the entire text and the pattern you are looking for will be inside a highlighted square, but we can combine other functions from the library with a *regexp*.

You can quickly find numbers, letters, punctuation, or spaces in texts. Look at the following code. The comments will show you what is being done:

```
# Find numbers
str_view_all(txt, "[:digit:]")
# Find letters
str_extract_all(txt, "[:alpha:]")
# Find punctuation
str_view_all(txt, "[:punct:]")
# Find spaces
str_view_all(txt, "[:space:]")
```

As a result, the code displays the following, where the square around a pattern is what the code has matched as the result of the *regexp*:

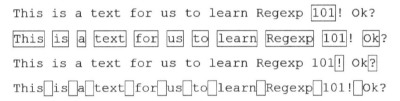

To build our own *regexp*, there are a few codes to learn. The following exercises consider the following text: *This is a text for us to learn Regexp 101! Ok?*. The function used is str_extract_all(), except where you see *, where the function used was str_view_all(). The squares in the text denote what the *regexp* code matches in the string:

What to Find	Pattern to use	Result
Specific Character	"[a]" or "a"	"This is a text for us to learn Regexp 101! Ok?"
Not match specific characters i or s*	"[^is]"	"This is a text for us to learn Regexp 101! Ok?"
Any single lowercase character from a to z	"[a-z]"	"This is a text for us to learn Regexp 101! Ok?"
Any single UPPERCASE character from A to Z	"[A-Z]"	"This is a text for us to learn Regexp 101! Ok?"
Any character – first match	"."	"This is a text for us to learn Regexp 101! Ok?"
Any single number within range 0 to 9	"[0-9]"	"This is a text for us to learn Regexp 101! Ok?"
Digits character*	"\\d"	"This is a text for us to learn Regexp 101! Ok?"
Not digits character*	"\\D"	"This is a text for us to learn Regexp 101! Ok?"
Letter character*	"\\w"	"This is a text for us to learn Regexp 101! Ok?"
Not letters character*	"\\W"	"This is a text for us to learn Regexp 101! Ok?"
White space character*	"\\s"	"This is a text for us to learn Regexp 101! Ok?"
Not white space character*	"\\S"	"This is a text for us to learn Regexp 101! Ok?"
Begins with	"^T"	"This is a text for us to learn Regexp 101! Ok?"
Ends with	"\\?$"	"This is a text for us to learn Regexp 101! Ok?"
Text between boundaries	"\\btext\\b"	"This is a text for us to learn Regexp 101! Ok?"
Pattern happens zero or more times	"[0*]"	"This is a text for us to learn Regexp 101! Ok?"
Pattern happens one or more times	"1+"	"This is a text for us to learn Regexp 101! Ok?"

* using str view all().

Let's practice these patterns a little more by referring to the book *Alice's Adventures in Wonderland*. The exercises that follow show us how versatile regexps are.

The following code will count how many times the main character's name was used in the book. Then, we will look only for digits. The third line of code checks for numbers written in long form. The next two lines count words in uppercase and those ending with the `ing` pattern:

```
# Exact Match of Alice
sum(str_count(alice$text, "Alice"))
[1] 399


# How many digits in the text?
sum(str_count(alice$text, "[:digit:]"))
[1] 2


# How many written numbers 1-5. | means "or".
sum(str_count(alice$text, "one|two|three|four|five"))
[1] 273


# How many words in UPPERCASE
sum(str_count(alice$text, "\\b[A-Z]+\\b"))
[1] 622


# All the words ending in "ing"
sum(str_count(alice$text, "\\b(\\w+ing)"))
[1] 975
```

Apparently, the author does not use numbers in digital form and likes uppercase. The words ending with ing are mostly gerunds and usually denote the act of doing something, according to thesaurus. com (https://www.thesaurus.com/e/grammar/whats-a-gerund/) – here, it's a character doing something. There are almost 1,000 (975) patterns of this kind, as seen in the returned value of the last snippet. Certainly, there are patterns that are not gerunds, such as nothing or king, that were captured, but we are not worried about that right now.

There is so much more that these expressions can do, but the main concepts have been covered here. Remember that you can always combine the functions learned about in the first section of the chapter with *regexp* patterns.

Next, we will combine the str_extract_all() function with the pattern for extracting the words ending in ing. Then, we will use the unlist() function to transform the result into a vector, as seen next:

```
# Extract the words ending in "ing"
gerunds <- c( str_extract_all(alice$text, "\\b(\\w+ing)") )
# Show only the values
unlist(gerunds)
```

```
 [1] "beginning"      "sitting"        "having"        "nothing"
        "reading"       "considering"
 [7] "making"         "getting"        "picking"       "nothing"
        "burning"       "considering"
[13] "stopping"       "falling"        "going"         "coming"
        "anything"       "killing"
[19] "nothing"        "tumbling"       "anything"      "getting"
        "thing"          "showing"
[25] "listening"      "_curtseying"    "falling"       "asking"
        "nothing"        "talking"
[31] "saying"         "dozing"         "walking"       "saying"
(...)
```

Regexps were created to be customizable and fit any needs to match patterns in text. Thus, I encourage you to practice and keep learning more. There are good resources for you to explore in the *Further reading* section. Likewise, when you feel it is too difficult to generate a *regexp* and the patterns don't match, remember that pages such as **Stack Overflow** are a valuable *go-to* resource.

Text is also data. One way or another, we can transform text into numbers, such as by counting the frequencies of words, which is one of the next section's exercises.

Creating frequency data summaries in R

With the concepts we have studied so far, I believe that you will already have a sense of the relevance of a frequency table. As a data scientist, you will be expected to create a frequency table on many occasions to gain a greater understanding of data and to know what is happening more frequently.

We already know that histograms and bar plots are great ways to visualize frequencies, but graphics do not always show numbers – due to lack of space or just because they have not been set up that way. As such, a frequency table can come handy, especially if you want to see the top 5 or top 10 items with a little more background than just the pure graphic.

Even though we can create frequency tables for numeric variables, it is not very common – nor useful, depending on the amount of data – because you will be counting a lot of unique observations. Frequency tables are more related to categorical data; therefore, we are covering them in the chapter about strings.

Let's see how regexps can be used on a daily basis with a practical example.

Regexps in practice

We will use Lewis Carrol's classic *Alice's Adventures in Wonderland* again to go over an exercise to wrap up this chapter.

The following code snippet begins using the `stringr` library with a *regexp* to extract all the words that end with `ing` in the book. `\\b` denotes a boundary, `(\\w+ing)` refers to letter characters, `+` is for one or more occurrences, and we have the `ing` pattern. After that, we will use `unlist()` on the object created to make it a vector of words. Now, if we use the `table()` function from base R, we can see each unique value and how many times it appears. On top of that, we transform the result of the `table()` function into a `data.frame` format. Finally, we order the results in descending order by frequency and collect the top 10 most frequently occuring ones:

```
# Extract the "ing" ending words
ings <- str_extract_all(alice, "\\b(\\w+ing)")
# Transform the words in a vector
ings <- unlist(ings)
# Data Frequency to data.frame
df_ings <- data.frame( table(ings) )
# Sorting descending and collect top 10 most frequent
observations
df_ings_top10 <- df_ings[order(-df_ings$Freq)[1:10],]
```

▲	ings	Freq ⇕
249	thing	80
116	King	62
129	looking	32
145	nothing	30
94	going	27
91	getting	22
6	anything	19
18	being	19
224	something	17
245	talking	16

Figure 4.1 – Frequency table of the 10 most frequently used words
ending with "ing" in Alice's Adventures in Wonderland

Certainly, we could do much more work on the table in *Figure 4.1*, such as only filtering for verbs, for example, but that is a task more related to NLP, which goes beyond the scope of this book.

In the next exercise, let's create a more complete table. To start, we will list the main characters from the book. Then, we will count how many times each of them appears in the story and place the results into a frequency table. Within the `characters` variable, we concatenate the names of the characters to look for. Then, using `str_c()`, we put all of them into a single textual pattern separated by `|`, which means *or* for *regexps*. We transform the names into Title Case for a better match, and use `str_extract_all()` combined with `unlist()` to get the vector for every match. Finally, we use `table()` to count every name and put the result into a `data.frame` object:

```
# List characters
characters <- c("Alice", "Rabbit", "Queen", "King", "Cheshire
Cat", "Duchess", "Caterpillar", "Hatter")
# Create regexp string
char_regexp <- str_c(characters, collapse = "|")
# Make the entire text Title Case for better match
alice_title_case <- str_to_title(alice$text)
# Extract the words
count_book_chars <- unlist( str_extract_all(alice_title_case,
char_regexp) )
# Data Frequency to data.frame
df_book_chars <- data.frame( table(count_book_chars) )
```

The result is the table in *Figure 4.2*:

	count_book_chars	Freq
1	Alice	399
7	Queen	77
6	King	64
5	Hatter	57
8	Rabbit	54
4	Duchess	42
2	Caterpillar	29
3	Cheshire Cat	6

Figure 4.2 – Frequency of the appearance of each character in the book

Now, we can go a little further and calculate the percentages, using the `proportions()` function and the cumulative percentage sum, using `cumsum()`, for each row. Initially, we create a variable with the data frame from the previous code snippet, `df_book_chars`, ordering it by the `Freq` variable. Then, we create a new column named `pct` by using the `proportions()` function to calculate the proportions, combined with the `round()` function to round them up to the nearest three decimal points. Finally, we create another column, `pct_cumSum()`, to hold the cumulative sum of the percentages calculated by the `cumsum()` function:

```
# Sort table by frequency
df_book_chars <- df_book_chars[order(-df_book_chars$Freq),]
# Add percentages rounded to 3 decimals
df_book_chars$pct <- round( proportions(df_book_chars$Freq), 3)
# Add cumulative sum of the pct
df_book_chars$pct_cumSum <- cumsum(df_book_chars$pct)
```

	count_book_chars	Freq	pct	pct_cumSum
1	Alice	399	0.548	0.548
7	Queen	77	0.106	0.654
6	King	64	0.088	0.742
5	Hatter	57	0.078	0.820
8	Rabbit	54	0.074	0.894
4	Duchess	42	0.058	0.952
2	Caterpillar	29	0.040	0.992
3	Cheshire Cat	6	0.008	1.000

Figure 4.3 – A more complete summary table

Figure 4.3 displays the results of our code. We can quickly see that Alice and the Queen of Hearts are the characters that appear the most in the story, which makes sense if you have ever read the book or watched one of the movie adaptions of it. Alice is our *hero* and the Queen is our *anti-hero* in this story. Another interesting insight captured here is that five characters make up 90% of the appearances in the story.

In programming languages, there is often more than one way to do something. We have just created a frequency table using our recent acquired knowledge about `stringr`, but there are other libraries that can do so, such as *gmodels*, which we will present next.

Creating a contingency table using gmodels

There is a library named `gmodels` that can be used to create good contingency tables too. First, you should install the package and load it into your RStudio session:

```
install.packages("gmodels")
library(gmodels)
```

The main function to create the contingency table is `CrossTable()`, a powerful, single line of code that brings up a good amount of information. To process this function, I will load the `mtcars` dataset one more time so that you can see what a contingency table created out of categorical numeric data looks like. As we have already discussed, a variable being numeric does not necessarily mean that it is a number used for mathematical operations. It could be a category, also known as a factor type in R, such as the number of cylinders or gears in a car:

```
#Load the data
data("mtcars")
```

```
# Create the contingency table
CrossTable(mtcars$cyl, mtcars$gear, prop.t=TRUE, prop.r=TRUE,
prop.c=TRUE)
```

As we can see in *Figure 4.4*, the table is complete – all you need is a single line of code with the CrossTable() function. The parameters are the variables you are *crossing*, prop.t will show the proportion of the cell count against the total from the table, prop.r shows the proportion against the total rows, and prop.c is the cell count against the total columns:

```
Cell Contents
|-----------------------|
|                     N |
| Chi-square contribution |
|           N / Row Total |
|           N / Col Total |
|         N / Table Total |
|-----------------------|


Total Observations in Table:  32
```

mtcars$cyl	mtcars$gear 3	4	5	Row Total
4	1	8	2	11
	3.350	3.640	0.046	
	0.091	0.727	0.182	0.344
	0.067	0.667	0.400	
	0.031	0.250	0.062	
6	2	4	1	7
	0.500	0.720	0.008	
	0.286	0.571	0.143	0.219
	0.133	0.333	0.200	
	0.062	0.125	0.031	
8	12	0	2	14
	4.505	5.250	0.016	
	0.857	0.000	0.143	0.438
	0.800	0.000	0.400	
	0.375	0.000	0.062	
Column Total	15	12	5	32
	0.469	0.375	0.156	

Figure 4.4 – Contingency table (output of gmodels)

There we have it – a nice table created with just a little coding, inviting you to constantly look for better ways to perform a task using R.

Text mining

NLP is a wide and active field of study. You probably interact with NLP algorithms at least once a day while using voice assistants, translators, or maybe speech-to-text converter. As you may have already noticed in this chapter, there is a lot of information to be extracted from a text, especially when we can count values and measure the importance of words.

For this section, we will use the `tidytext` library, which helps us to mine text very easily with just a few functions. Before we dive in to the following subsections, let's load a dataset for our examples. It is the book *The Time Machine*, by H. G. Wells, downloaded from the open source `gutenberg` library for R:

```
# Downloading "The Time Machine" by H. G Wells
book <- gutenberg_download(gutenberg_id = 35)
```

Let's move on.

Tokenization

We should start with the definition of a token. A token is the smallest meaningful unit of a text. It is most common to find words as tokens, but a token can also be a phrase, a sentence, or any other unit that makes sense for the project. To illustrate the concept and build our intuition around it, let's consider this example phrase: *He is a data scientist with many skills.* If we tokenize this sentence by word, we will get *he, is, a, data, scientist, with, many, skills.* Therefore, one word becomes one token.

But after tokenization, despite how tokens are supposed to be the smallest unit from a text that carries some meaning, we can end up with many meaningless tokens that are used to connect words and form phrases, like prepositions, conjunctions, some verbs. Good examples are *the, is, a, with*, and *many*. These meaningless units are called **stop words** in text mining, and they are usually removed from the dataset during the tokenization step, before the analysis, otherwise we would run the risk of getting results that suggest that the words *the, is*, and *a* are the most important ones in the text, which is far from true.

We already know that a **tidy** dataset is easier to work with, as there will be a single observation per row and a single variable per column. When we are talking about text mining, the concept of **tidiness** is defined as *one token per row*.

Let's perform some tokenization. The code is quite simple. We call our `book` dataset and use the pipe symbol to connect it to the following task, `%>%`, which is the `unnest_tokens(output= "tokens", input= text)` function. The function takes the output argument, which determines the name of the column to hold the tokens, and the second argument is the input column from our dataset. Notice that we are not assigning the result to any variable, so the code will be only for visualizing the table:

```
# Tokenization
```

```
book %>%
  unnest_tokens(output= "tokens", text)
```

The code will yield the table in *Figure 4.5*.

A tibble: 32,761 x 2

gutenberg_id	tokens
<int>	<chr>
35	the
35	morlocks
35	x
35	when
35	night
35	came
35	xi
35	the
35	palace
35	of

41-50 of 32,761 rows

Figure 4.5 – Tokenized dataset. One token per row

It is easy to see that there are many stop words in the output, even in this small extract presented in *Figure 4.5*. Therefore, next we will clean it with another simple function. Let's use unnest_tokens() combined with anti_join(). So, the code will take the dataset, tokenize it, and blend it with stop_words, but in this case, we use anti_join so we keep only whatever is not in the stop_words dataset. The argument to make the connection between the datasets is by= c('tokens' = 'word'), where we indicate which columns contain the words to be matched:

```
# Tokenization and clean stop words
clean_tokens <- book %>%
  unnest_tokens(output='tokens', input= text) %>%
  anti_join(stop_words, by= c('tokens' = 'word'))
```

If we look at the clean data now, this is what we see.

gutenberg_id <int>	tokens <chr>
35	returns
35	iv
35	time
35	travelling
35	golden
35	age
35	vi
35	sunset
35	mankind
35	vii

11-20 of 11,268 rows   Previous   1   2

Figure 4.6 – Dataset tokenized and cleaned of stop words

As you can see in *Figure 4.6*, the dataset is now clean. More refined cleaning could be done to remove words such as the Roman numbers *vi* and *vii*, but let's stick with what we've got for now.

To close this exercise, let's make a word frequency counter. We can just use the object from the last step, `clean_tokens`, and chain it with the `count()` function, offering the `tokens` variable as input:

```
# Counting words frequency
clean_tokens %>%
  count(tokens, sort=TRUE)
```

The result is shown here:

A tibble: 4,172 x 2

tokens <chr>	n <int>
time	207
machine	88
white	61
traveller	57
hand	49
morlocks	48
people	46
weena	46
found	44
light	43

1-10 of 4,172 rows   1   2

Figure 4.7 – Most frequent words in The Time Machine

Interesting (and kind of expected!) result. The book *The Time Machine* has the tokens *time* and *machine* as the top two most frequently mentioned words.

Now, moving on to the next concept, we will learn more about stemming and lemmatization.

Stemming and lemmatization

Stemming uses the stem (root) of a word to reduce the number of variations of a word that occur in an analysis. This way, word variations such as *say* and *saying* will be reduced to its root *sai* – stems don't always make sense but can still be useful. To perform stemming using **tidytext** is simple. We will use the clean_tokens object once again, and then we will use the mutate() function to create a new variable named stem for the new words. With that done, we use count() to get the frequencies:

```
# Stemming
clean_tokens %>%
  mutate(stem = wordStem(tokens)) %>%
  count(stem, sort = TRUE)
```

The preceding code's results are shown in *Figure 4.8*.

stem <chr>	n <int>
time	220
machin	92
travel	90
hand	77
white	61
dark	57
morlock	51
light	49
peopl	46
weena	46

1-10 of 3,190 rows

Figure 4.8 – Most frequent words in The Time Machine, after stemming

Lemmatization, on the other hand, uses a word's lemma, that is, the word's basic meaning. The lemma takes into account the context in which the word is being used. Hence, the words *say* and *saying* will become simply *say*, which is the basic meaning of both variations.

To perform this task, we will need to load other libraries: **tm** and **textstem**. First, we will create a custom function to help us clean the text (you can see the code in the GitHub repository). This function takes in the text from the book, transforms it into a corpus object, and cleans it, removing punctuation, whitespaces, and stop words. Just as a side note, a corpus is a type of textual object composed of a collection of textual documents and metadata, commonly used in text mining projects.

The next step is to apply it to clean the text:

```
# Clean text
clean_text <- clean_corpus(book$text)
```

Then we can perform the lemmatization. In the following code, the tm_map() function takes in clean_text and adds the lemmatize_strings function, saving the result in clean_lem. Then, we take this object and transform it into a matrix of terms by documents, where each row of the text is considered a document and how many times each word is in that document is counted. We do that using the TermDocumentMatrix function. Next, we transform that object into a regular matrix, TDM. Finally, we will count the word frequencies and sort, until we are ready to transform the result into a data.frame object:

```
# Lemmatization
clean_lem <- tm_map(clean_text, lemmatize_strings)
TDM <- TermDocumentMatrix(clean_lem)
TDM <- as.matrix(TDM)
word_frequency <- sort(rowSums(TDM),decreasing=TRUE)
lemm_df <- data.frame(word = names(word_frequency),freq=word_
frequency)
```

Once that is done, if you open the lemm_df data frame, you will notice that the algorithm could not get rid of the " and ", so we can do that manually:

```
# Remove word == " or "
lemm_df <- lemm_df %>% filter(!word %in% c('"', '"') )

# View head
head(lemm_df, n=10)
```

Figure 4.9 shows the resulting table.

word <chr>	freq <dbl>
time	203
come	136
little	112
upon	104
say	90
thing	90
machine	86
think	74
seem	74
hand	70

Figure 4.9 – Word count of The Time Machine after lemmatization

Observe that machine now drops to 7th position. The word *come* is in the second place now, probably because of variations such as *coming* and *came*. Therefore, this shows us that, in this case of word frequency, stemming and lemmatization are not necessarily the best options if you want to get a sense of a book's content. In our first exercise with a simple word count, the words *time, machine, morlocks, traveler,* and *weena* appeared in the top 10, and they make a lot more sense if you know this book, because those words are related to the theme of the book and main characters.

But there are still two other interesting ways to extract good information from textual data. Next, we will study Term Frequency-Inverse Document Frequency.

TF-IDF

TF-IDF stands for Term Frequency-Inverse Document Frequency. This measure is used regularly in text mining to weigh the importance of words. The rationale behind this metric is that if a word appears many times in a document inside a corpus (like a chapter in a book), it will be more important that another word that appears many times throughout the entire corpus. Let's illustrate the concept.

Consider the following corpus with three documents, in this case three phrases:

- (d1) *My dog is my beautiful dog.*

- (d2) *My dog is an amazing animal.*

- (d3) *My friend has a pet.*

The **Term Frequency** (**TF**) will measure how many times a token appears in a document. The calculation of the TF for each word within a document is as simple as [**TF**] / [**Total terms**]. So, the TF for *dog* in d1 is $2/6 = 0.33$, in d2 it is $1/6 = 0.17$, and it is $0/4 = 0$ in d3. Therefore, the word `dog` is more frequent in d1 than in any other document.

The **Inverse Document Frequency (IDF)** measures how important a term is. It calculates a logarithm of the inverted frequency. So, **log([Number Documents]/ N Docs with the term])**. This calculation will lower in importance terms that are very frequent in every document and will scale up terms that are more rare. The IDF for dog is `log(3/2) = 0.4054`.

To calculate the final importance of the word for each document, we just multiply the measures [**TF**] * [**IDF**]. In d1, TF-IDF for *dog* is `0.33 * 0.4054 = 0.133782`; in d2, `0.068918`; and in d3, `0`.

If we consider only the TF, notice that the word *my* will be considered the most important, having `4/17` appearances against `3/17` of *dog*. Using TF-IDF, *my* will always have a score of 0, since it appears in all the documents and `log(3/3) = 0`, compared to the previous calculated scores for *dog*. Indeed, the word *dog* brings more meaning to the text. Now let's see how to perform that calculation in RStudio.

We will continue using the same dataset, so let's jump into the coding. The `clean_tokens` object is our dataset, tokenized and cleaned of all stop words. Using the `mutate()` function, we will add an index column, `idx`:

```
# Add column with the indexes
tokens_by_chapter <- clean_tokens %>%
  mutate(idx = 1:nrow(clean_tokens)) %>%
  select(idx, tokens)
```

Now, we will add a column with chapter numbers using the `mutate()` function, combined with `case_when()`, which takes the the `idx` column and determines the chapter to assign to a given observation based on the numbers of the index:

```
# Adding the chapter column
tokens_by_chapter <- tokens_by_chapter %>%
  mutate(chapter =
           case_when(
             idx < 621 ~ "introduction",
             between(idx, 621, 1045) ~ "ii",
             between(idx, 1045, 1764) ~ "iii",
             between(idx, 1764, 2566) ~ "iv",
             between(idx, 2566, 3143) ~ "v",
             between(idx, 3143, 3930) ~ "vi",
             between(idx, 3930, 4695) ~ "vii",
             between(idx, 4695, 6098) ~ "viii",
             between(idx, 6098, 6825) ~ "ix",
             between(idx, 6825, 7670) ~ "x",
             between(idx, 7670, 8509) ~ "xi",
```

```
                    between(idx, 8509, 9407) ~ "xii",
                    between(idx, 9407, 9777) ~ "xiii",
                    between(idx, 9777, 10537) ~ "xiv",
                    between(idx, 10537, 10700) ~ "xv",
                    between(idx, 10700, 11171) ~ "xvi",
                    idx >= 11171 ~ "epilogue"      )    ) %>%
        select(chapter, tokens)
```

The next code snippet is used to count how many times a token repeats by chapter. So, we use the count() function to count the words by chapter:

```
# Words by chapter
tokens_by_chapter <- tokens_by_chapter %>%
  count(tokens, chapter, sort=T)
```

Next, we use bind_tf_idf(term= tokens, document= chapter, n= n) to add the TF-IDF calculations:

```
# TF-IDF calculation
data_tf_idf <- tokens_by_chapter %>%
  bind_tf_idf(tokens, chapter, n)
```

Finally, we can plot this using the **ggplot2** library. We will learn more about the library in Part 3. For now, just follow the code. Starting from the data_tf_idf object, we group it by chapter and use slice_max() to get only the largest n=5 items ordered by the tf_idf variable. The with_ties = F argument allows only exactly five items per chapter. Then we use ungroup() to get rid of grouped levels. The result is stored in plot_tf_idf:

```
# Data to plot
plot_tf_idf <- data_tf_idf %>%
  group_by(chapter) %>%
  slice_max(tf_idf, n = 5, with_ties = F) %>%
  ungroup()
```

To plot the result, we chain plot_tf_idf with the ggplot() function, pass x, y, and fill colors by chapter. Then, we add the geom_col() graphic type and facet_wrap() to create many plots, one by chapter divided in four columns, and use the tf_idf numbers as the x axis labels:

```
#Plot TF-IDF
plot_tf_idf %>%
  ggplot(aes(x=tf_idf, y=fct_reorder(tokens, tf_idf), fill =
```

```
chapter)) +
  geom_col(show.legend = FALSE) +
  facet_wrap(~chapter, ncol = 4, scales = "free") +
  labs(x = "tf-idf", y = NULL)
```

This is the result, divided into two parts in *Figure 4.10*.

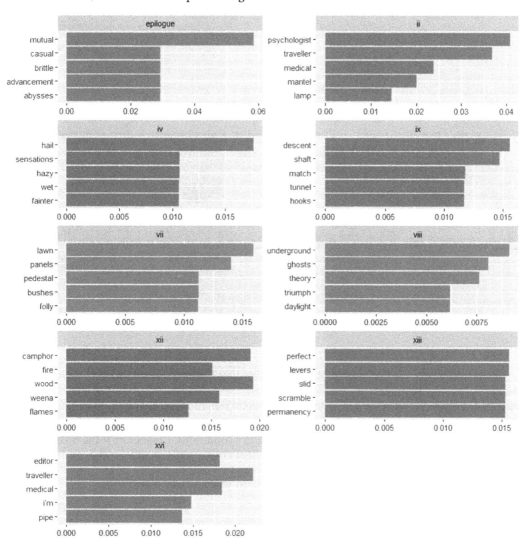

Figure 4.10 – Top five tokens by chapter after TF-IDF calculations

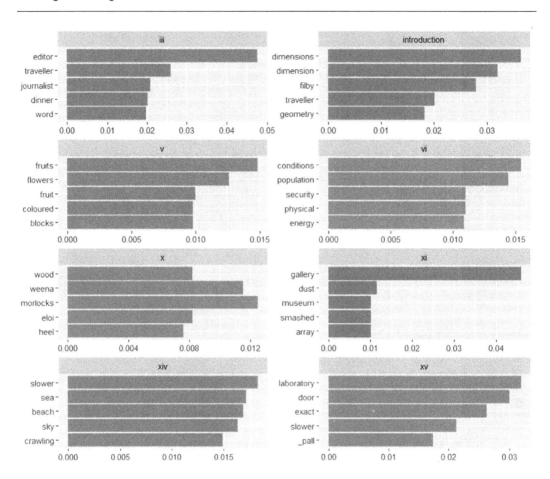

Figure 4.11 – Top five tokens by chapter after TF-IDF calculations (continued)

Look how the character *traveller* appears a lot in chapters *introduction*, *ii*, and *iii*. In chapter *vi*, for example, the character travels through time and reaches a future point in time, where he describes how things are, talking about the population and security conditions. Chapter *xv* is the final one, so it is when the character is at the laboratory revisiting his journey. It is a good job from TF-IDF.

Next, let's learn how to understand words in their context when analyzed in groups.

N-grams

N-grams are groups of N words in sequence. Text mining projects tend to work with bi-grams or tri-grams, which are groups of two and three words in sequence, respectively. Those small sequences are preferred because they are more likely to repeat. A sequence of four words or more is much less likely to happen several times. From the sentence *The trip was an adventure*, we can extract the following bi-grams: *the trip, trip was, was an*, and *an adventure*. Naturally, before we extract the N-grams, it is a good idea to clean the text, so we eliminate stop words.

Let's look at the most used bi-grams from our dataset. The unnest_tokens() function already has a handy argument to manage that. In the next code, we will use our raw data and extract the most common two words in sequence. For that, we will just add the token= "ngrams", n=2 argument to the code:

```
# Most frequent 2-grams
book %>%
  unnest_tokens(output = ngrams,
                input = text,
                token= 'ngrams', n=2) %>%
  count(ngrams, sort=T)
```

The result can be seen in *Figure 4.11*.

ngrams <chr>	n <int>
NA	456
of the	291
in the	161
i had	120
i was	107
and the	101
the time	97
it was	95
to the	82
as i	78

1-10 of 18,631 rows

Figure 4.12 – Most common bi-grams from the book

We observe that it is flooded with meaningless words, but I wanted to show this so that you know the importance of cleaning the text before analysis. Let's use the clean corpus now and extract the same table to compare the result. First, we get the clean_tokens object to make the dataset a single text again. We will select only the tokens column and use the str_c() function from **stringr** to concatenate the words, storing it in a tibble object named clean_text:

```
# Gather clean text
clean_text <- clean_tokens %>%
  select(tokens) %>%
  str_c(sep=" ", collapse = NULL) %>%
  as.tibble()
```

Next, we can re-run the same code we used for the raw data from the book:

```
# Most frequent 2-grams
clean_text %>%
  unnest_tokens(output = ngrams,
                input = value,
                token= 'ngrams', n=2) %>%
  count(ngrams, sort=T)
```

The result is displayed next.

ngrams <chr>	n <int>
time traveller	56
time machine	40
white sphinx	12
green porcelain	10
palace green	10
time travelling	9
time traveller's	8
looked round	7
golden age	5
hundred thousand	5

1-10 of 10,790 rows

Figure 4.13 – Most common bi-grams from the book's clean text

What a difference. *Figure 4.12* shows us a very different picture, with bi-grams that are much more related to the book's content.

Finally, let's learn about categorical variables, what R calls factors. They are also an important string type.

Factors

Since we are talking about strings, it is important to mention factors in R. These are objects used to create categorical variables, whether those categories are ordered or not. Categories can be text or numbers, for instance. We can create a factor variable using factor():

```
# Textual variable
var <- c('A', 'B', 'B', 'C', 'A', 'C')
```

```
# To create a factor
factor_var <- factor(var)
```

And, subsequently, you can see what the environment shows:

```
values
   factor_var              Factor w/ 3 levels "A","B","C": 1 2 2 3 1 3
   var                     chr [1:6] "A" "B" "B" "C" "A" "C"
```

Figure 4.14 – Text variable and factor variable

As seen in *Figure 4.5*, the levels have been created and the order assigned is alphabetical. If we want to change that, we can use `levels()`:

```
# Ordered levels
levels(factor_var) <- c('C','B','A')
factor_var


[1] C B B A C A
Levels: C B A
```

If we print the variable again after changing the order, see how 1 2 2 3 1 3 now becomes C B B A C A, confirming that the levels have changed.

Factors are interesting objects because they already code variables *under the hood*. Some algorithms like to receive variables as numbers, so if their type is factor, this works too. It's variable encoding for the computer, keeping the text that is quicker for us humans to understand.

We have closed our study of strings, the first major variable type. In the next chapter, we will learn more about numbers, another common type.

Summary

This chapter brought in many tools for working with strings. To refresh our memory, string are textual information. Like any other data, text can carry a lot of information and good insights if we have the right tools to extract it. That was exactly what we did in the last few pages.

We learned about the main functions from the `stringr` library, which is part of the tidyverse package. The benefit of learning how to use libraries from tidyverse is easy coding and its adherence to the tidy concept, helping you transform your data and prepare it for modeling. We saw that `stringr` functions start with the `str_` prefix followed by a task identifier, making it easier to learn and remember the coding.

Next, we covered the most used regexp patterns, which are strings for creating highly customized textual patterns for many different analysis tasks.

Finally, we learned how to use a combination of `stringr` functions plus regexps and some base R functions, to create data summaries, as well as presenting the `gmodels` library, which includes a single function for creating very complete contingency tables that can enhance your analysis. We covered factor types at the end of the chapter since they are a well-known type of text used for categorical data in R.

Exercises

1.   What is a string?

2.   What is the difference between the `print()` and `writeLines()` functions?

3.   What function can be used to find a string?

4.   What function can be used to find the index of a pattern in a text?

5.   What is the function used to count the number of repetitions of a string?

6.   How can we concatenate strings?

7.   What is a regexp?

Further reading

- StringR official documentation:

  `https://stringr.tidyverse.org/`

  `https://github.com/rstudio/cheatsheets/blob/main/strings.pdf`

- Six tips about regexps:

  `https://tinyurl.com/3nachdfx`

- Practice using regexps by writing them and seeing the results in real time:

  `www.regexp101.com`

- Regex Generator:

  `https://regex-generator.olafneumann.org/`

5
Working with Numbers

Variables are quantitative when they quantify a measurement of something. Numbers are the representation of those measurements, which will most likely vary for each observation and start to create variation patterns that can tell us you a lot about a subject.

In this chapter, we will work with numbers, learning how to handle them in vectors, matrices, or data frames, since there are differences in terms of dimensions and functions available for each data type.

Once we have that covered, it is then time to see how to do math operations in RStudio, not only using basic functions but also creating custom functions, which we will apply to numbers, making our set of tools more powerful so we can deal with many kinds of problems.

When working with numbers, it is hard not to talk about descriptive statistics, such an important step of data exploration. Statistics such as average, median, percentiles, standard deviation, and correlation are all about identifying central points and how spread out our data points are. Those statistics can tell us a lot about data, leading our thinking toward better solutions.

In this chapter, we will cover the following main topics:

- Numbers in vectors, matrices, and data frames
- Math operations with variables
- Descriptive statistics

Technical requirements

All the code can be found in the book's GitHub repository: `https://github.com/PacktPublishing/Data-Wrangling-with-R/tree/main/Part2/Chapter5`.

Numbers in vectors, matrices, and data frames

A number represents a point in space. You may also have heard of a number being referred to as a scalar when it is followed by a unit of measure. In other words, it is a variable with a single number. When we have more than one number, it is possible to create a line in space, which is referred to as a vector. A collection of vectors put together gives new dimensions to data, which becomes matrices or data frames. These last two are similar structures, but data frames have some more enhanced features, such as headers and indexes, that help us to work with the information held by them.

We can quickly go over scalar, vector, matrix, and data frame creation in R, which is a simple process. You can understand what is being done by reading the comments:

```
# Creating a scalar
scalar <- 42
print(scalar)
[1] 42


# Creating a vector
vec <- c(1, 2, 3, 4, 5, 6, 7, 8, 9)
print(vec)
[1] 1 2 3 4 5 6 7 8 9


# Creating a Matrix
mtrx <- matrix(data=vec, nrow = 3, ncol= 3)
print(mtrx)
      [,1] [,2] [,3]
[1,]    1    4    7
[2,]    2    5    8
[3,]    3    6    9


# Creating Data frame
df <- data.frame(column1= c(1,2,3),
                 column2= c(4,5,6),
                 column3= c(7,8,9))
print(df)
```

column1 <dbl>	column2 <dbl>	column3 <dbl>
1	4	7
2	5	8
3	6	9

3 rows

Figure 5.1 – A data frame in R, which has the same data as a matrix
but with additional features for data handling

After creating one of each of those objects, let's look closer at each of them.

Vectors

Vectors are collections of elements, such as a list of numbers, but I don't like to call them *lists* because, in R, a list is a wholly different object. When a vector of numbers is created, R will assign an index to each number, starting with one up to the length of the vector, in a way that makes it easier to find every single element from that vector. As shown in *Figure 5.2*, the index is related to the position of the element within the vector, rather than to its value. Such a configuration is useful for extracting parts of vectors, matrices, or data frames for data wrangling, in a task we call *slicing*.

The full dataset is not always needed. There are many examples to illustrate the reason why we would slice a dataset, such as extracting only a filtered part of the data to replace a value, performing a mathematical operation for a group, or removing certain values of a dataset. Therefore, slicing can be extremely important for data wrangling to help us filter the data.

In the vector in *Figure 5.2*, slicing the first three elements would give us 9, 8, and 7.

Figure 5.2 – R indexing for vectors – each element can be found by an index number

Look at the code that creates the vector from *Figure 5.2* and then slices the first three elements:

```
# Vector
vec <- c(9,8,7,6,5,4,3,2,1)

# Slicing a vector
vec[1:3]
[1] 9 8 7
```

Note that the indexing in R starts with 1 (some programming languages start with 0) and the last value of the slicing is inclusive.

Performing operations with vectors reveals that they are elementwise, meaning that the operations with objects with the same length will result in a new vector created from operations with elements in the same index position. By adding vectors, for example, they will go index 1 + index 1, index 2 + index 2, and so on. When one of the vectors is smaller than the other, the smaller one is recycled; therefore, it goes back to the first item once the end is reached.

Let us create two new vectors:

```
# Vector
vec1 <- c(9,8,7,6,5,4,3,2,1)
vec2 <- c(1,2,3,4,5,6,7,8,9)
```

Next, we can slice vec1 with the condition that only numbers greater than or equal to 6 will remain.

```
# Slice by condition
vec1[vec1 >= 6]
[1] 9 8 7 6
```

We can sum these vectors. As they have the same length, the sum is elementwise 9 + 1, 8 + 2, 7 + 3 and so on, resulting in a bunch of 10s:

```
# Sum of vectors
vec1 + vec2
[1] 10 10 10 10 10 10 10 10 10
```

The same is true for multiplication:

```
# Multiplication of vectors
vec1 * vec2
[1]  9 16 21 24 25 24 21 16  9
```

If we multiply or add the vector to a constant number, the operation is the constant by each element of the vector:

```
# Addition or Multiplication of a vector by a single number
vec1 * 10
[1] 90 80 70 60 50 40 30 20 10
vec1 + 10
[1] 19 18 17 16 15 14 13 12 11
```

Finally, this is the behavior of operations with vectors of different sizes, when the smaller vector is recycled. Note that the additions are 1 + 1, 2 + 2, 1 + 3, and 2 + 4.

```
# Two vectors of different sizes
c(1,2) + c(1, 2, 3, 4)
 [1]  2  4  4  6
```

Speaking the data science language, a vector can be understood as a variable, a collection of measurements about a given attribute or quality. Scaling up, if we gather some vectors, we will have a matrix, which is the next object we will look at.

Matrices

The addition brought by matrices in comparison to vectors is the second data dimension. Vectors are unidimensional, while matrices have two dimensions – rows and columns. That addition requires a slightly different notation to slice data. You must now state the row and column position of the element within square brackets. Observe it in the following code:

```
# Create a Matrix
mtrx <- matrix(1:12, nrow=4, ncol=3, byrow=T)


# Slicing [row,col]
mtrx[2,3]
 [1]  6
```

In *Figure 5.3*, the slicing notation (in red) is added to the table to show which slicing will return each value. From the preceding code, we use [2,3], returning the expected number 6 as output.

	V1	V2	V3
1	[1,1] 1	[1,2] 2	[1,3] 3
2	[2,1] 4	[2,2] 5	[2,3] 6
3	[3,1] 7	[3,2] 8	[3,3] 9
4	[4,1]10	[4,2]11	[4,3]12

Figure 5.3 – R indexing for matrices – slicing requires [row, col]

We will now move on to data frames because they are very similar to matrices, having the same 2D configuration and, on top of that, having interesting features, such as named rows and columns and different data types, in the same object.

Data frames

Data frames are one of the most enhanced ways to store and work with data. We can get a little fancier if we work with Tibbles (see *Chapter 2*), but data frames will do well most of the time. There are many libraries and algorithms that expect a data frame as input, given its qualities.

The data frame's best qualities are the named rows and columns, making it even easier to index, and they are heterogeneous, which means that there could be variables of different types. In data frames, we can join text, numbers, and factors in the same object. Let's first create a data frame:

```
# Data frame
df <- data.frame(name = c('Carl', 'Vanessa', 'Hanna',
'Barbara'),
                 class_ = c('Math', 'Math', 'Math', 'Math'),
                 grade = c(8.5, 9, 9, 7)    )
```

Next, we can start slicing it by row and column number.

```
# Slicing row 1 and col 3 (Carl's grade)
df[1,3]
[1] 8.5
```

It is possible to slice by variable name too, using it between quotes instead of the column number:

```
# Slicing by variable
df[, "grade" ]
[1]   8.5  9.0 9.0  7.0
```

And we can slice based on a logical test, a condition, as seen in the next code snippet, which will return what is shown in *Figure 5.4*.

```
# Slicing by condition
df[ df$grade > 8,]
```

	name	class_	grade
	\<chr>	\<chr>	\<dbl>
1	Carl	Math	8.5
2	Vanessa	Math	9.0
3	Hanna	Math	9.0

Figure 5.4 – A data frame sliced using condition for the > 8 grade variable

You may have noticed that there are some slicing notations with an empty space before or after the comma. This means that we want to display all the rows or columns. For example, `df[df$grade > 8,]` has a condition that slices the rows where the grade must be above 8 and the column's space is blank, denoting that we want to see all the columns associated with that condition.

Since vectors, matrices, and data frames are no longer strangers to us, we can move on to creating operations using variables.

Math operations with variables

As part of a data wrangling process, there will be tasks involving mathematical operations with variables, where there will be a need to add, multiply, or even calculate the log of numbers, for example. Ergo, working with a data frame or a Tibble object is recommended, due to the facilities to perform those operations with variables.

The most common math operators in R are as follows:

Operator	Operation	Example
+	Addition	10 + 10 = 20
-	Subtraction	10 - 8 = 2
*	Multiplication	10 * 10 = 100
/	Division	9 / 3 = 3
% / %	Integer Division: returns only the integer	8 % / % 3 = 2. 8/3 - 2.667, but it returns only the integer.
^ or **	Integer Exponent: returns only the integer	2 ^ 3 = 8 (2 to the power of 3).
% %	Modulus: returns the remainder after the division	17 %% 4, then the result = 1.

Figure 5.5 – A table with the R language's math operators

If we still use the data frame with names and grades, just created for the last exercise, let's imagine that the professor offered one extra point for those who wrote a paper. Let's suppose everyone delivered it; here is how we can add a new column with the extra point:

```
# Extra point
# Scenario: everyone delivered
df$new_grade = df$grade + 1
```

name	class_	grade	new_grade
<chr>	<chr>	<dbl>	<dbl>
Carl	Math	8.5	9.5
Vanessa	Math	9.0	10.0
Hanna	Math	9.0	10.0
Barbara	Math	7.0	8.0

4 rows

Figure 5.6 – One point added to all the students

If the professor wants to normalize the grade column, where the highest number would be given the full grade (100%) and the other grades would be a percentage of that maximum, it can be done this way:

```
# Normalization
max_grade <- max(df$grade)
df$grade <- df$grade/max_grade
```

The result is shown in *Figure 5.7*.

name	class_	grade
<chr>	<chr>	<dbl>
Carl	Math	0.9444444
Vanessa	Math	1.0000000
Hanna	Math	1.0000000
Barbara	Math	0.7777778

4 rows

Figure 5.7 – The normalized grade column

Carrying out operations with variables in data frames is not that different than carrying out operations with vectors. At the end of the day, we already learned that each variable is, in essence, like a vector. Therefore, the math operations will have similar behavior. Therefore, I won't spend much time describing each operation, but I will leave this general syntax rule: to perform mathematical operations with variables, just point to the variables and the operation desired, such as var 1 *operation* var 2. Next, there are some examples of code for different operations with variables from a data frame:

```
# Perform math operation with variables
df$var1 + df$var2
df$var1 * df$var2
(df$var1 + df$var2) / df$var3
log(df$var)
df$var * 100
```

Even though operations with variables are fairly common and similar to operations with vectors, sometimes we need to repeat the same function over and over, for each element of a vector, list, or data frame. In those cases, creating a loop structure can solve the problem, but it generally does not result in the best performance, especially for the very large datasets that are common these days.

Fortunately, there is a family of built-in functions in R that helps us solve four different variations of this problem – this is the *apply* family.

apply functions

Another useful resource built in with the R language is the family of `apply()`, `sapply()`, `lapply()`, and `tapply()` functions. The logic behind them all is equivalent, which is to apply a certain function to a group of elements, such as a vector, list, or data frame.

For the next exercises, we will keep using the `grades` dataset created and modified earlier in this chapter. Just to refresh our minds, the table is presented again in *Figure 5.8*.

name	class_	grade	new_grade
<chr>	<chr>	<dbl>	<dbl>
Carl	Math	8.5	9.5
Vanessa	Math	9.0	10.0
Hanna	Math	9.0	10.0
Barbara	Math	7.0	8.0

Figure 5.8 – The grades dataset with the new_grade variable added

apply

The first function to be presented is the function that names the family – apply(). This lets you apply a function to the rows or columns of a data frame. Note that there is a slicing notation filtering only the numerical columns, leaving out the name and class_ variables, which are strings and can't have the mean() function applied to them. The function calculates the average of grade and new_grade for every row. MARGIN=1 means that we are applying the function by row:

```
# apply function sum to rows: sum 'grade' + 'new_grade'
apply(df[,c('grade', 'new_grade')], MARGIN= 1, FUN = mean)
[1] 9.0 9.5 9.5 7.5
```

In the second snippet, we used the same notation but sliced the data with numbers to select the columns instead, and calculated one mean for each variable. MARGIN=2 means that we are applying the function by column.

```
# apply function mean to columns the column 3 and 4
apply(df[,c(3,4)], MARGIN= 2, FUN= mean)
    grade new_grade
    8.375     9.375
```

Finally, we can apply a custom function as well. It is defined with function(x) {...}, and we use it as the FUN argument:

```
# apply custom function
my_func <- function (x){sum(x)/2}
apply(df[,c(3,4)], MARGIN= 1, FUN= my_func)
[1] 9.0 9.5 9.5 7.5
```

Note that my_func is the mean between grade (column 3) and new_grade (column 4).

lapply – list apply

Moving to the next one, the lapply() function is used to apply a function to lists of objects, returning a list object of the same length. Remember that lists are a collection of objects in R, being able to hold a combination of vectors, data frames, and graphics as each of its elements. Once again, we defined a custom function and applied it to the new_grade column using lapply():

```
# lapply of a custom function to remove the extra point from
new_grade
my_func <- function (x){x-1}
lapply_obj <- lapply(df$new_grade, my_func)
```

Name	Type	Value
🔽 lapply_obj	list [4]	List of length 4
[[1]]	double [1]	8.5
[[2]]	double [1]	9
[[3]]	double [1]	9
[[4]]	double [1]	7

Figure 5.9 – lapply() returns a list object

Figure 5.9 confirms that the resultant object of the `lapply()` function is a list.

sapply – same length apply

The `sapply()` function will also apply one single function to each object, but this time it returns an array or matrix object of the same length as result. Let us create a custom function once more and use it as input for `sapply()`:

```
# sapply of a custom function to a single column
my_func <- function (x){x-1}
sapply(df$new_grade, my_func)
[1] 8.5 9.0 9.0 7.0
```

We can also apply my_func to two columns at once:

```
# sapply of a custom function to two columns
sapply(df[,c(3,4)], my_func)
```

As expected, it returns an object with the same length, thus two columns:

```
      grade new_grade
[1,]    7.5       8.5
[2,]    8.0       9.0
```

```
[3,]    8.0        9.0
[4,]    6.0        7.0
```

The returned values are interesting. When we used `sapply()` on a single column, it returned an array. When we applied it to two columns, it returned a matrix with two columns. Therefore, it returns an object of the same length as the input, as the apply function documentation says.

tapply – text apply

The last function of this powerful family is the `tapply()` function. This one applies your function to each factor variable of a vector. In our example, let's add another class, `Math2`, to the data frame of *Figure 5.8* and then compute the mean for each class using `tapply()`:

```
# Create new df
# Data frame
df <- data.frame(name = c('Carl', 'Vanessa', 'Hanna',
'Barbara', 'Jason', 'Alison', 'Kevin', 'Melody'),
                 class_ = c('Math', 'Math', 'Math', 'Math',
'Math2', 'Math2', 'Math2', 'Math2'),
                 grade = c(8.5, 9, 9, 7, 5, 7, 10, 9.5)    )
```

To calculate the mean for each class, `Math` and `Math2`, we use `tapply()` and input the `grade` values, the index, which is the `class_` names, and the `mean` function to be applied to each group:

```
# Calculate the mean for each class_
tapply(df$grade, df$class_, mean)
Math Math2
8.375 7.875
```

This function is a quick way to group by a variable and apply a function to it. In this example, we grouped by classes (`Math` and `Math2`) and applied the `mean` function to each group, receiving the result that the `Math` class is doing better than `Math2` in terms of grades.

We saw that this family of functions is versatile, helping us to repeat the same function on all the elements of an object, so it is definitely a good addition to any data scientist's toolbox.

Working with numbers and not relating them to statistics is not easy. I am sure that right now, many statistical measurements rapidly come to our minds, such as mean, median, percentile, and correlation. Furthermore, R is software created for statistical computing, so it has a number of built-in functions to make these calculations, unlike other programming languages, where you must load specific modules for that purpose. Knowing this, nothing is more logical than making use of the core of R language, which is what we are about to do.

Descriptive statistics

Data is everywhere. So, when a dataset is created, it can be understood as a subset of a larger amount of data. Imagine a sales report of the last quarter, or a dataset with ages and heights of elementary students in a county, or even responses to an election poll. All of them are subsets of a larger universe of data. Let's think about that for a minute – the sales report does not show all the history of sales, the ages and heights are not for all students across the country, and the election poll does not contain responses from every citizen eligible to vote. Hence, these are examples of **samples**, which were collected from the whole, which is called the **population**.

The population holds the true values of mean, median, maximum, and minimum, and when we refer to these metrics in relation to the population, they are called **parameters**. If it was possible to have all the data and there was enough computational power to process it, we could just use those values for data analysis. However, we know that this scenario is not true. It does not even make sense to collect all the data for analysis. It is a waste of time and resources, but that is where statistics come to play.

Now, think of a good hot bowl of soup. You think it is delicious, but you are not sure. What do you do? Do you eat all the soup to taste it, or do you take a *sample* with a spoon to taste it to get an idea of how good the soup really is? I assume you chose the second option. And that is what statistics is all about – making inferences about the population using data from the sample. When you calculate those same measurements as mean, median, maximum, and so on, but now using the values of the **sample**, these numbers are called **statistics**.

Statistics can be calculated in RStudio with the built-in functions exemplified as follows. Let's create two random variables with 50 numbers in a normal distribution, with a mean of 0 and standard deviations of 1 and 5:

```
# Creating a variable: 50 data points, mean 0, standard
deviation 1
var1 <- rnorm(50, mean = 0, sd=1)
# Creating a variable: 50 data points, mean 0, standard
deviation 10
var2 <- rnorm(50, mean = 0, sd=5)
```

Let's calculate the mean value:

```
# Mean
mean(var1)
mean(var2)
```

Let's calculate the standard deviation:

```
# Standard Deviation
sd(var1)
sd(var2)
```

Let's calculate the median:

```
# Median
median(var1)
median(var2)
```

Let's calculate the 25%, 50% (median), and 75% quartiles:

```
# percentiles or quantiles
quantiles(var1, c(0.25, 0.5, 0.75))
quantiles(var2, c(0.15, 0.55, 0.90))
```

The results are not displayed here, but you can see them in the R script related to this chapter on GitHub (`https://tinyurl.com/zf6s3jbm`).

Statistics from the sample help data scientists describe the way a variable is spread in space. Each statistic will help them gain knowledge about a characteristic of the variable by summarizing it as a single number. This is why we call it descriptive statistics, since we are describing the variable with numbers.

The following code snippet creates a figure with two columns to plot (`par(mfrow=c(1,2))`) and then plots two histograms for each of those random variables previously created:

```
# Setup grid with 2 columns
par(mfrow=c(1,2))

# Histogram var1
hist(var1, col='royalblue', ylim=c(0,20), xlim=c(-20,20) )
# Histogram var2
hist(var2, col='coral', ylim=c(0,20), xlim=c(-20,20) )
```

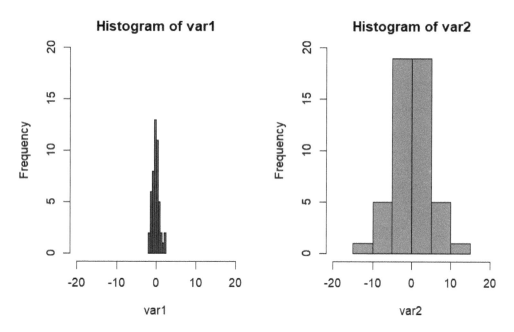

Figure 5.10 – Comparing how spread out two variables are

Figure 5.10 helps us visualize the spread of the two variables. Both were created within the same *X* and *Y* axis limits, so the comparison is natural. Note how tight `var1` is and how spread `var2` is. Now, let's describe both variables statistically.

We can create a data frame with `var1` and `var2`:

```
# Create a data frame with both variables for comparison
df <- data.frame(var1= var1,
                 var2= var2)
```

Then, using the recently learned `apply` function, we will calculate the mean, standard deviation, minimum value, 25%, 50%, and 75% quartiles, and the maximum value by each column. Let's see how we did that:

```
# Comparison data frame
data.frame( avg= apply(df, 2, mean),
            std_dev= apply(df, 2, sd),
            min_val= apply(df, 2, min),
            pct25= apply(df, 2, function (x)
{quantile(x,0.25)}),
            median_val= apply(df, 2, median),
```

```
            pct75= apply(df, 2, function (x)
{quantile(x,0.75)}),
            max_val= apply(df, 2, max)
            )
```

The code will result in the data frame shown in *Figure 5.11*.

| | avg | std_dev | min_val | pct25 | median_val | pct25 | max_val |
	<dbl>	<dbl>	<dbl>	<dbl>	<dbl>	<dbl>	<dbl>
var 1	-0.1429296	0.8663997	-1.997642	-0.7591558	-0.26984044	0.3566766	2.072036
var 2	0.4029616	4.2894005	-10.746300	-2.0902128	-0.04780342	4.0470340	10.101674

2 rows

Figure 5.11 – Descriptive statistics of the variables

With *Figure 5.10* in mind, the plan is to compare it to the numbers returned as a data frame in *Figure 5.11*. Both variables have their mean near 0, and the median is also close to the mean. As the minimum and maximum values are almost symmetrical and the mean is a central value, we can sense that the data is nearly normal.

From the minimum, median, maximum, and quartile values, we can also get an idea of how spread the data is. While `var1` jumps 0.489 from the 25th percentile to the median and 0.626 from the median to the 75th percentile, `var2` moves 2.043 and 4.096 respectively; therefore, there is much more spread.

The result is satisfactory, since we were able to analyze the statistics of the variables, but that required some coding. As we learned before, R generally has quicker or simpler ways to perform the same task. In this case, another way to calculate descriptive statistics in R is by using the built-in `summary()` function:

```
# Descriptive stats with summary function
summary(df)


      var1                 var2
 Min.    :-1.9976   Min.    :-10.7463
 1st Qu.:-0.7592    1st Qu.: -2.0902
 Median :-0.2698    Median : -0.0478
 Mean    :-0.1429   Mean    :  0.4030
 3rd Qu.: 0.3567    3rd Qu.:  4.0470
 Max.    : 2.0720   Max.    : 10.1017
```

In the preceding code, note that it returns the 0, 25, 50, 75, and 100 percentiles. However, it lacks important measurements, such as the mean value and the standard deviation. However, we can call a fancier function, this time from the `skimr` library. You can call it as follows:

```
library(skimr)
skim(df)
```

A tibble: 2 x 11

skim_variable	n_missing	complete_rate	mean	sd	p0	p25	p50	p75	p100
<chr>	<int>	<dbl>	<dbl>	<dbl>	<dbl>	<dbl>	<dbl>	<dbl>	<dbl>
1 var 1	0	1	-0.1429296	0.8663997	-1.997642	-0.7591558	-0.26984044	-0.3566766	2.072036
2 var 2	0	1	-10.746300	4.2894005	-10.746300	-2.0902128	-0.04780342	4.0470340	10.101674

2 rows | 1-11 of 11 columns

Figure 5.12 – Descriptive statistics from the skimr library

The resultant table in *Figure 5.12* resembles our personally created data frame from *Figure 5.11*, and it brings even more information, such as the number of missing values.

Calculating descriptive statistics is an important step during data exploration. It has so much information about a dataset and the way it is distributed, making it possible to draw preliminary insights that will drive other exploration efforts for wrangling and modeling. Knowing whether you are working with a normal or a skewed distribution drives decisions such as using the mean or the median as the central value measurement.

For symmetric distributions such as the normal one, the mean and standard deviation make a great duo by providing the central position and the spread. On the other hand, if you have a skewed distribution, you probably will see more outliers that will distort the mean value; thus, it would not the best option to take as a central reference. The median is more suitable in this case, since it is a statistic robust against outliers.

A good statistical test to make when working with numbers is the correlation test, which we will see next.

Correlation

Correlation is a statistical test to measure the strength of the linear relationship between two numerical variables. There are three commonly used tests (Pearson, Spearman, and Kendall), with Pearson's being the most famous. It was created by Karl Pearson in the 1880s and is based on the Pearson correlation coefficient, which is basically a normalization of the covariance by the standard deviation. In plain English, it tells us whether the A and B variables are above or below their averages, at the same time giving us information on whether the variables are in the same direction or opposite each other.

Correlation results in a number between -1 and 1. The closer this number is to the extremes, 1 or -1, the stronger the relationship is between the two variables. The closer it is to 0, the weaker the relationship is. A simple way to understand the result of the test is if when A increases and B also increases in a similar proportion, the relationship is positive, so you shall see numbers close to 1. When A increases but B decreases in a similar proportion, the number will get closer to -1. If they don't influence each other, the result will be closer to 0.

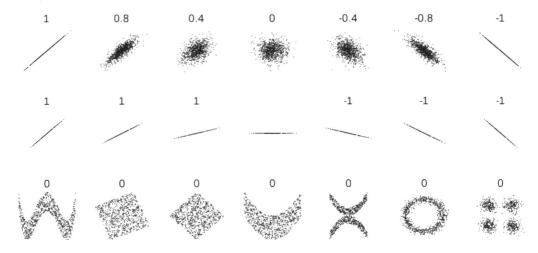

Figure 5.13 – Correlation and examples of data

When performing the test in R, the code brings another built-in function, cor(). Next, we will see how to calculate the correlation between our example variables, var1 and var2:

```
# Correlation
cor(var1, var2, method='pearson')
[1] -0.09196411
```

We can plot a scatterplot to see the relationship between them:

```
# Scatterplot
plot(var1, var2, col='royalblue', pch=16)
```

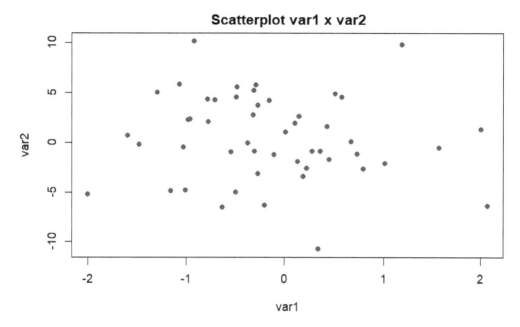

Figure 5.14 – A correlation plot of var1 x var2

Revisiting what we have learned, if the correlation measures the strength of a linear relationship between two variables and the scatterplot does not bring any linear relationship at all, it is expected that the result of the test will be close to zero (-0.0919).

This concludes this chapter, where we learned a lot about numbers in R and how to make good use of functions to handle them during data analysis.

Summary

In this chapter, numbers were on display. The R language is great for dealing with numbers, since the software was created as a statistical tool. As we know, statistics is all about numbers, so we were able to see that many of the functions used during this chapter are from the Base R, eliminating the need to install or load any library to work with so many useful functions.

We started the chapter by learning about structures with numbers, such as vectors, matrices, and data frames. That knowledge prepared us for the next section, where we studied many operations to deal with numbers in vectors and data frames, and we learned a good resource for that is the `apply` family of functions.

We also went over how descriptive statistics are important to help us gain an understanding of data and its distribution, because that can drive our efforts of data wrangling before modeling.

Finally, we saw the correlation test and how to interpret its result.

Exercises

- What are a vector, a matrix, and a data frame?

- What is the difference between matrices and data frames?

- What is data slicing and why is it important for data wrangling?

- List the four functions of the apply family.

- List three descriptive statistics functions.

- How can you display a statistics summary in R?

- What does it mean when a correlation is close to 1 and when it is close to 0?

Further reading

- Collection of functions apply: `https://tinyurl.com/4h8km3c9`

- How to Fix in R: dim(X) must have a positive length.

- If you try to use the `apply()` function in a single column, you might see this error. In this link you can read more about how to fix it.

- `https://tinyurl.com/yc2ffh45`

- Dealing with numbers in R: `https://tinyurl.com/2p98p7ns`

- Code for this chapter in GitHub: `https://tinyurl.com/2nm6ev48`

6
Working with Date and Time Objects

Computers are complex machines that can, among many other things, record date and time. The world is so attached to the measurement of time that it drives a great part of our daily lives. We work within a time interval, we are supposed to go to school for a determinate minimum number of years, and even the value of our work is calculated based on time. Likewise, the business world is also run by the wheel of time. This is evident because of the popular saying: *time is money*.

Dates and times are just like other types of data, holding valuable information and good business insights if you know how to handle them. During a data exploration that involves a `datetime` variable, questions such as *What is the busiest month, day, and hour? Which days of the week have more traffic?* or *How much was the revenue in the past 3 months?* will arise, and we must know how to deal with those objects to facilitate better analysis.

There are three main ways to work with datetime objects in Data Science:

- Creating them
- Manipulating existent datetime objects
- Arithmetic operations with datetime

We will go over these three ways of working during this chapter. For that purpose, we will use **lubridate**, another library that composes the **tidyverse** package. To be ready to code along with this chapter, use the `library(lubridate)` command to load it to your RStudio session. I believe that, by the end of this chapter, we will agree about the relevance of studying date and time objects for data wrangling.

We will cover the following main topics:

- Introduction to date and time
- Date and time with lubridate
- Date and time using regular expressions (regexps)
- Practicing

Technical requirements

Dataset: We will use the Classic Rock dataset, from FiveThirtyEight. To access the original dataset, go to `https://github.com/fivethirtyeight/data/tree/master/classic-rock`

All the code can be found in the book's GitHub repository: `https://github.com/PacktPublishing/Data-Wrangling-with-R/tree/main/Part2/Chapter6`.

Here are the libraries to be used in this chapter:

```
library(tidyverse)
library(lubridate)
```

Introduction to date and time

In R, there are three objects related to date and time: `date`, `time`, and `datetime`. The definition is as logical as it looks:

- **date**: Object refers to a date `YYYY-MM-DD`
- **time**: Object stores time data, as `HH: MM: SS`
- **datetime**: A combination of both `YYYY-MM-DD HH: MM: SS`

From now on, I will mostly refer to *datetime* objects, as these are a combination of both other types and are the most common as well.

It is important to note that computers calculate time based on **January 1, 1970, at 00:00:00** UTC. As a side note, **UTC** means **Universal Time Coordinated**, formerly known as **Greenwich Mean Time (GMT)**, which is the point that regulates the world time zones. Every calculation of time zone is done from that point, adding or subtracting hours.

In the late 1960s and early 1970s, Unix engineers had to pick a date to use as *ground zero* when the clock started to count for computers. For the sake of calculation easiness, January 1 was convenient, and also there would not be much going on with computers before that date. This is also referred to as the Unix Epoch (in `https://www.compuhoy.com/when-did-unix-time-start/`).

Having said that, we now know that every date calculation will start from `1970-01-01` and will be counted in seconds from that date and time. If we create a date object at the zero mark, we can see as follows:

```
as_date(0)
[1] "1970-01-01"
```

`lubridate` has nice features, such as the capability to easily pull date and time from the system where it is running, with the `today()` and `now()` functions, as demonstrated next:

```
# Checking today's system date
today()
[1] "2022-07-14"
# Checking current system's time
now()
[1] "2022-07-14 22:33:50 EDT"
```

The function is so precise that it even provides time zone information, identifying that the system is currently on **Daylight Saving Time (DST)** on the eastern coast of the USA.

To create datetime objects with `lubridate`, it is necessary to know the meaning of each letter associated with a determinate time period. Then, it is just a matter of choosing the datetime format you like the best or the one required by your project. If you don't have a preference or your project does not impose any specific format, consider using the universal format—that is, **year-month-day-hour-minute-second**.

Look at *Figure 6.1* to gain more knowledge about the `lubridate` letters pattern used in functions to create datetime objects with this library:

Datetime period	Associated letter
Year	y
Month	m
Day	d
Hour	h
Minute	m
Second	s

Figure 6.1 – Letters associated with each period of date and time

It is basically the first letter of each datetime period used in lowercase. The next step to create an object is just a matter of gathering those letters in the order you need the dates and times to be. So, if you want to create a **year-month-day** format, use the ymd() function. For **day-month-year**, use dmy(). In the same way, to create a datetime object, just separate the date and time portions with an underscore (_), using the ymd_hms() function.

See the following code examples and their respective returned values:

```
# Creating a date object
ymd(20220714)
[1] "2022-07-14"
mdy('Jul142022')
[1] "2022-07-14"
# Using year and quarter information
yq('2010Q4')
[1] "2010-10-01"
# Creating a time object
hms::as_hms(43200) #(12 hours*60min*60sec=43200sec)
12:00:00
ymd_hms(20220714150000)
[1] "2022-07-14 15:00:00 UTC"
# Creating a datetime object one year after 1970-01-1
#(60s*60min*24h*365d = 31,536,000 seconds)
as_datetime(31536000)
[1] "1971-01-01 UTC"
```

A couple of things to point out: the output of a datetime created is always in the universal format, and the standard to calculate time is always in seconds—that is why 43200 becomes 12 hours in the fourth function presented.

If a variable is presented in datetime format but, for any reason, there is a reason to drop the time portion of the data, that can be done with the code that follows. We can create a datetime object with ymd_hm() and then assign it as a date object with as_date():

```
# Datetime object
dt_tm <- ymd_hm( "2022-01-02 03:04" )
[1] "2022-01-02 03:04:00 UTC"
# Convert to just date object
as_date( dt_tm)
[1] "2022-01-02"
```

Observe that it dropped the time information, keeping only the date. Not all projects will need granularity to the minute. Then, a single line of code is enough to perform the task elegantly.

There still is a lot of work ahead of us. We already know how to create datetime objects with `lubridate`, being that important in many ways for data science projects, such as using input from people or from a system to create a variable, for example. Let's learn how to manipulate them when they already exist.

Date and time with lubridate

Dates and times have their formatting as the main characteristic to distinguish this type of data. A quick look at the variables where a YYYY-MM-DD number appears is enough to tell that it is a date object. However, as mentioned, computers calculate date and time based on seconds, so it is not difficult to see a dataset that brings a variable *date* or *time* as an integer number. In those cases, the solution is to recur to the data dictionary (document with the description of each variable) or to the dataset owner and align if that column should indeed be treated as a datetime object or a regular number. Later in this chapter, we will see this problem in action and how to solve it.

Before that, let's set the base by learning some fundamental functions that will help us to parse datetime objects, splitting them into separate objects. Once again, I will ask you to go over the table from *Figure 6.2* to get familiar with the logic of the `lubridate` library for cases such as this. For this exercise, consider the date of `"2000-01-02 03:04:05"`, written in the universal format, and assigned to the variable name `dt`. The value to be returned is what is highlighted in red:

Period to extract	Function	Result
Year	`year(dt)`	2000-01-02 03:04:05
Month	`month(dt)`	2000-01-02 03:04:05
Day	`day(dt)`	2000-01-02 03:04:05
Hour	`hour(dt)`	2000-01-02 03:04:05
Minute	`minute(dt)`	2000-01-02 03:04:05
Second	`second(dt)`	2000-01-02 03:04:05
Week	`week(dt)`	1
Weekday	`wday(dt)`	1 (Sunday)
Time zone	`tz(dt)`	"UTC"

Figure 6.2 – Parsing function associated with each period of date and time

The `lubridate` library makes it so intuitive that parsing datetime in R becomes a simple task. Of course, it can be more complicated than that, since each project is singular, but at least the basic parsing should be smooth for you after this chapter.

In the code from *Figure 6.2*, we could just replace the `dt` object with a vector of dates or a variable from a data frame, and the result would be the same. So, you have the intuition built. Use it whenever needed.

Supposing there are three variables with strings of year, month, and day that we need to join in a single date object, first it is necessary to put those strings in a date sequence and then call the adequate function. In the next code snippet, we can add the values of our date to separate variable names:

```
# Separate variables
y_obj <- "2022"
m_obj <- "5"
d_obj <- "10"
```

After that, we can gather them using the `ymd()` function combined with the `paste()` function to gather the three parts and create a date object:

```
# gather date
ymd( paste(y_obj,m_obj,d_obj, sep="-") )
[1] "2022-05-10"
```

In the same way it worked for these strings, it works for data frames too. The parsing from `lubridate` is quite powerful and holds many formats built into it, therefore it recognizes many datetime patterns. In the example we just saw, the code works either with or without the `sep` parameter from the `paste()` function. In the *Further reading* section, you will find a link to the library's documentation, where there are many examples of different datetime strings that can be parsed using this library.

The sequence brings arithmetic operations with datetime. Let's dive into it.

Arithmetic operations with datetime

Arithmetic operations with datetime are the third bullet among the main wranglings done with date and time. They are, though, slightly more complex than operations with real numbers. The complexity comes from different periods and specific properties, such as time zones. If we add 20 days to January 10, we get to the 30th. But if we add it to February, we will land in March, for example.

Look at this example, where we subtract `dt2` from `dt1`:

```
# Simple subtraction
dt1 <- as_date("2022-06-01")
dt2 <- as_date("2022-05-01")
dt1 - dt2
Time difference of 31 days
```

Subtracting them makes sense. `dt1` happened after `dt2`. But if I try to add both dates instead of subtracting, it will not make sense to the computer to sum two dates, and it will throw an error: `Error in `+.Date`(dt1, dt2) : binary + is not defined for "Date" objects`. It says that the sum operation is not defined for dates, which is logical since it is difficult to know which numbers to sum in that operation. We can sum the number of days to a date, but not a date with another one. That simple example illustrates how datetime operations can be tricky.

Life is surrounded by additions and subtractions of time: *How long since I arrived here? How long until the warranty period ends? How many days until Christmas?* These are just a few questions we pose and hear every day. So, to comprehend operations with date and time, we must understand *periods*, *durations*, and *intervals*:

- **Period**: Tracks changes in time, not accounting for any deviations, such as gaps. It facilitates the addition and subtraction operations with time.

- **Duration**: Tracks the passage of time, accounting for deviations.

- **Interval**: A time interval composed of a start and an end.

Period

A **period**, in `lubridate`, will assist R in calculating the sum and subtraction of datetime. Again, operations with date and time are not like math with other numerical types. Therefore, the creation of a **period** comes in handy when there is a need to add or subtract time from a given date. To create this type of object, we use the name of the time period desired for the operation in the plural form: `years()`, `months()`, `days()`, `minutes()`, and so on.

As an example, in the case of machinery maintenance, one can use a period to know the exact date of the next checkup. Once you have a datetime object and want to add time to it, you can create a **period** and then add or subtract it easily with `date + period` or `date - period`. Here is the code used in this example. We create a `date` object:

```
# Date of the last maintenance
dt <- ymd("2021-01-15")
```

Then we create a period (p) object, composed of a combination of `years()`, `months()`, and `days()`:

```
# Create a period to add or subtract
p <- years(x=1) + months(x=06) + days(x=1)
# Another syntax
p <- period( c(1, 6, 1), c("year","month", "day"))
```

Next, there is the calculation, wrapped with the `writeLines()` function and also using the `paste()` function to gather a text followed by the calculation, providing us with a nicer output:

```
# Calculation
writeLines( paste("Next maintenance date is on:", dt + p) )
Next maintenance date is on: 2022-07-16
```

Keeping in mind that computers use seconds as their standard time unit for calculations, then it makes more sense that every time we create a **duration**, that will be calculated in seconds. To create an object like that, use the letter d followed by the time period desired in the plural form: `dyears()`, `ddays()`, and `dseconds()`, for example. The code for creating a duration of 5 years is as follows:

```
# Create a duration of 5 years
dw <- dyears(x=5)
dw
[1] "157788000s (~5 years)"
```

Duration

The **duration** can be useful to calculate the ending date of a warranty, for example. We create a date object, first, with `ymd()`:

```
# Date
dt <- ymd("2000-01-01")
```

Then, we create a 5 years duration object with `dyears()`:

```
# Create a duration of 5 years
dw <- dyears(x=5)
dw
```

And the calculation can be, once again, calculated and printed with some additional text:

```
# Calculate warranty time 5 years after dt
warranty_end <- dt+dw
writeLines( paste("Warranty ends on:", warranty_end) )
Warranty ends on: 2004-12-31 06:00:00
```

Interval

Last, but not least, the **interval** is a bounded period of time, with start and end dates. It can be quite useful, among other things, to know if a datetime is within that interval.

Let us create a date object once more:

```
# Date
dt <- ymd("2022-01-01")
```

Create an interval object with the `interval()` function, inputting the start and end dates:

```
# Interval start
i <- interval(start= "2021-01-01", end= "2022-12-31")
i
[1] 2021-01-01 UTC--2022-12-31 UTC
```

And we can test if the date (`dt`) is within the interval (`i`) or not:

```
# Date within interval
dt %within% i
[1] TRUE
```

In the preceding code snippet, January 1, 2022 is within the time interval tested, thus returning TRUE.

As the world these days is very connected, time zones are an important resource to manage time in different parts of the globe. Let's see more about that next.

Time zones

We already saw one function that shows the time zone: `tz()`. But there are other interesting topics about that matter. As a good start, observe that `lubridate` can get the system's time zone:

```
# System time zone
Sys.timezone()
[1] "America/New_York"
```

A result to be expected from the previous code could have been something such as **Eastern Standard Time (EST)**, for example. Let's imagine a time in **Central Standard Time**, known as **CST**, in the USA, although CST could also relate to time zones from other countries, such as Mexico, Honduras, or Nicaragua. A quick search in Google can tell you that (`https://tinyurl.com/s396a7t9`). For that reason, R will use the international standard **Internet Assigned Numbers Authority (IANA)** time zones naming convention, which shows the continent/city, just like the `America/New_York` result from the last code snippet.

For datetime object creation, the library will work with the UTC zone by standard, unless you specify otherwise using the `tz` parameter in the datetime creation functions. In the next code block, you can see a standard datetime and a customized object:

```
# Creating a datetime object in another timezone
ymd_hms("2022-01-01 00:00:00")
[1] "2022-01-01 UTC"
ymd_hms("2022-01-01 00:00:00", tz="Europe/Paris")
[1] "2022-01-01 CET"
```

If you are curious about the time zone names, type `OlsonNames()` into your RStudio console to see a complete list.

Yet another quick transformation to do with dates is to display them in another time zone. Say you are in Dubai, but you need to see dates in New York's time zone. The code to do such a transformation is as follows:

```
# Date creation
dt_dubai <- ymd_hms("2022-07-01 10:00:00", tz="Asia/Dubai")
[1] "2022-07-01 10:00:00 +04"
with_tz(dt_dubai, tzone="America/New_York")
[1] "2022-07-01 02:00:00 EDT"
```

Time zones serve us the good purpose of organizing the time around the world, and that was well captured by `lubridate` when it comes to changing your object to the time on a different part of the planet.

We still have one topic left to study, which is how to use the regular expressions learned in *Chapter 4* to parse dates out of a text.

Date and time using regular expressions (regexps)

The datetime functions in `lubridate` can parse dates out of a good number of cases, even from phrases. Observe how the `mdy()` function can correctly parse only the date, which is in a weird format, by the way:

```
# Lubridate parsing
mdy("The championship starts on 10/11-2000")
[1] "2000-10-11"
```

But certainly, that feature combined with `regexp` is even more powerful. If we try to use the same `mdy()` function, this time we will get an error message: `Warning: All formats failed to parse. No formats found.` Regular expressions can pick every date from a text. Let's create an example text to help illustrate this exercise:

```
# Text
t <- "The movie was launched on 10/10/1980. It was a great hype
at that time, being the most watched movie on the weeks of
10/10/1980, 10/17/1980, 10/24/1980. Around ten years later, it
was chosen as the best picture of the decade. The cast received
the prize on 09/20/1990."
```

Next, using `str_extract_all()` from `stringr` and a `regexp` pattern, we can see the five dates in the output:

```
# Parse using regex
str_extract_all(t, "[0-9]+/[0-9]+/[0-9]+")


[[1]]
[1] "10/10/1980" "10/10/1980" "10/17/1980" "10/24/1980"
"09/20/1990"
```

Using a regular expression allows a lot of customization and the ability to parse all the dates from a text at once, saving time and effort. After all, as you know, *time is money*.

Practicing

Before starting this practice, we should understand that this exercise is good for us to know the possibilities of working with datetime objects. However, there are some functions and libraries that we still did not fully cover, so you might see new functions in this section. Don't worry. We will cover all of this in this book, and you can always come back to this chapter later to review the more challenging code.

Let's practice the use of datetime variables using a dataset from *FiveThirtyEight*, about classic rock. The dataset has observations of songs played in many radio stations in one week of June 2014, which we can use to gain some insights about that period in time.

The variables in this dataset are as follows:

- `SONG RAW`: Song title

`Song Clean`: Song title after cleaning up the name, removing not unmeaningful words such as *live*

`ARTIST RAW`: Artist name

`ARTIST CLEAN`: Artist name after removal of nonmeaningful elements and correcting name abbreviations

`CALLSIGN`: Station that played the song

`TIME`: The date and time the song was played

`UNIQUE_ID`: Unique ID assigned to each play

`COMBINED`: `Song Clean` and `ARTIST CLEAN` combined

`First?`: Binary column with `1` if it was the first mention of the song, or `0` if not

The dataset can be found at this link (`https://github.com/fivethirtyeight/data/tree/master/classic-rock`). Let's load it into RStudio:

```
# URL where the data is stored
url <- "https://raw.githubusercontent.com/fivethirtyeight/data/
master/classic-rock/classic-rock-raw-data.csv"

# Load to RStudio
df <- read_csv(url)
```

A screenshot of the data is in the sequence:

	SONG RAW	Song Clean	ARTIST RAW	ARTIST CLEAN	CALLSIGN	TIME	UNIQUE_ID	COMBINED	First?
1	Caught Up In (live)	Caught Up in You	.38 Special	.38 Special	KGLK	1402943314	KGLK1536	Caught Up in You by .38 Special	1
2	Caught Up In You	Caught Up in You	.38 Special	.38 Special	KGB	1403398735	KGB0260	Caught Up in You by .38 Special	0
3	Caught Up In You	Caught Up in You	.38 Special	.38 Special	KGB	1403243924	KGB0703	Caught Up in You by .38 Special	0
4	Caught Up in You	Caught Up in You	.38 Special	.38 Special	KGLK	1403470732	KGLK0036	Caught Up in You by .38 Special	0
5	Caught Up in You	Caught Up in You	.38 Special	.38 Special	KGLK	1403380737	KGLK0312	Caught Up in You by .38 Special	0
6	Caught Up in You	Caught Up in You	.38 Special	.38 Special	KGLK	1403105300	KGLK1162	Caught Up in You by .38 Special	0
7	Caught Up in You	Caught Up in You	.38 Special	.38 Special	KGLK	1402970932	KGLK1446	Caught Up in You by .38 Special	0

Figure 6.3 – TIME column has integers

If you look at the dataset loaded (*Figure 6.3*), the `TIME` column brings integer numbers. To be able to work with these dates and use the functions from `lubridate`, we must first transform that variable to a datetime variable. Let's use our recently acquired skills and apply the `as_datetime()` function to the `TIME` variable:

```
# Variable TIME to datetime
df$TIME <- as_datetime(df$TIME)
```

	SONG RAW	Song Clean	ARTIST RAW	ARTIST CLEAN	CALLSIGN	TIME	UNIQUE_ID	COMBINED	Fir
1	Caught Up In (live)	Caught Up in You	.38 Special	.38 Special	KGLK	2014-06-16 18:28:34	KGLK1536	Caught Up in You by .38 Special	
2	Caught Up In You	Caught Up in You	.38 Special	.38 Special	KGB	2014-06-22 00:58:55	KGB0260	Caught Up in You .38 Special	
3	Caught Up In You	Caught Up in You	.38 Special	.38 Special	KGB	2014-06-20 05:58:44	KGB0703	Caught Up in You .38 Special	
4	Caught Up in You	Caught Up in You	.38 Special	.38 Special	KGLK	2014-06-22 20:58:52	KGLK0036	Caught Up in You .38 Special	
5	Caught Up in You	Caught Up in You	.38 Special	.38 Special	KGLK	2014-06-21 19:58:57	KGLK0312	Caught Up in You .38 Special	
6	Caught Up in You	Caught Up in You	.38 Special	.38 Special	KGLK	2014-06-18 15:28:20	KGLK1162	Caught Up in You by .38 Special	
7	Caught Up in You	Caught Up in You	.38 Special	.38 Special	KGLK	2014-06-17 02:08:52	KGLK1446	Caught Up in You .38 Special	
8	Caught Up In You	Caught Up in You	.38 Special	.38 Special	KRFX	2014-06-22 16:58:23	KRFX0060	Caught Up in You by .38 Special	

Figure 6.4 – TIME column transformed to datetime

For our purpose of collecting insights about the data, there is no need for all the variables. Columns such as SONG RAW, ARTIST RAW, and UNIQUE_ID will not be useful. The information they carry is contained in other variables, thus let's select what matters to this exploration. See the code as follows:

```
# Select variables
df <- df[,c("Song Clean", "ARTIST CLEAN",
            "CALLSIGN", "TIME", "COMBINED",
            "First?")]
```

Next, we are creating some extra granularity for time, by adding columns for month, day, weekday, and hour for further analysis. This is important when the analyst needs to group data by day or by hour, for example, using different periods of time. To create the new columns, we can code as described in the following code block:

```
# Add new column year
df$year <- year(df$TIME)
# Add new column month
df$month <- month(df$TIME)
# Add new column day
df$day <- day(df$TIME)
# Add new column week day
df$weekday <- wday(df$TIME)
# Add new column hour
df$hour <- hour(df$TIME)
```

Notice in *Figure 6.5* that the year is always 2014 and the month is 06, so there's no point in keeping them if you don't want to. I am just adding them here for educational purposes:

Clean	ARTIST CLEAN	CALLSIGN	TIME	COMBINED	First?	year	month	day	weekday	hour
it Up in You	.38 Special	KGLK	2014-06-16 18:28:34	Caught Up in You by .38 Special	1	2014	6	16	2	18
it Up in You	.38 Special	KGB	2014-06-22 00:58:55	Caught Up in You by .38 Special	0	2014	6	22	1	0
it Up in You	.38 Special	KGB	2014-06-20 05:58:44	Caught Up in You by .38 Special	0	2014	6	20	6	5
it Up in You	.38 Special	KGLK	2014-06-22 20:58:52	Caught Up in You by .38 Special	0	2014	6	22	1	20
it Up in You	.38 Special	KGLK	2014-06-21 19:58:57	Caught Up in You by .38 Special	0	2014	6	21	7	19
it Up in You	.38 Special	KGLK	2014-06-18 15:28:20	Caught Up in You by .38 Special	0	2014	6	18	4	15
it Up in You	.38 Special	KGLK	2014-06-17 02:08:52	Caught Up in You by .38 Special	0	2014	6	17	3	2
it Up in You	.38 Special	KRFX	2014-06-22 16:58:23	Caught Up in You by .38 Special	0	2014	6	22	1	16

Figure 6.5 – New columns added for year, month, day, weekday, hour

Since now there are many slices of time, we can create some different views of the data. To begin, the question to be answered with data is: *What is the distribution of distinct songs played per day of the week?*

Breaking down the code, we first create a filtered dataset with unique observations:

```
# Filter only unique observations
df_unique <- df[!duplicated(df$`Song Clean`),]
```

The second step is to use the `table()` function to count occurrences by weekday:

```
# Songs by weekday
song_by_wkd <- table(df_unique$weekday)
```

Finally, let's visualize the result with a bar plot:

```
# Bar plot of Songs by weekday
barplot(song_by_wkd, col="royalblue",
        main="Number of unique songs by day of the week",
        xlab="Day of the Week [1=Sun, 7=Sat]",
        ylab= "Distinct Songs")
```

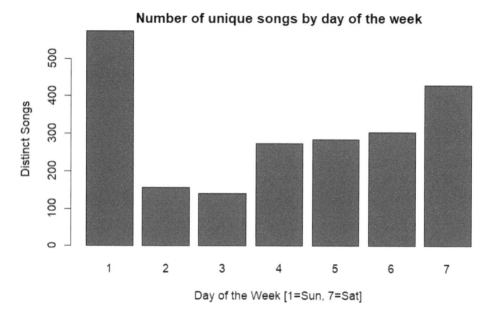

Figure 6.6 – Distribution of songs played during weekdays

From *Figure 6.6*, we can understand that the weekends have more distinct songs played. Tuesday is the worst day for those music lovers who want to discover new songs, as that is the day with the fewest different songs played.

The next question to be answered is: *What is the hour when more songs are being played?*

In this case, there is no need to filter only distinct songs, as we really want to know what the busiest hour of those days is, taking all the stations into consideration. To begin, we save the observations by hour in `song_by_hour`. Our data is tidy, so there is one observation per row. So, if we count how many occurrences of each hour number have happened, then we will know how many songs were played at that time:

```
# Songs by hour
song_by_hour <- table(df$hour)
```

Let's plot a bar plot to see the result:

```
# Bar plot of Songs by hour
barplot(song_by_hour, col="royalblue",
        main="Songs by hour",
        xlab="Hour",
        ylab= "Songs Played")
```

The output is shown in *Figure 6.7*:

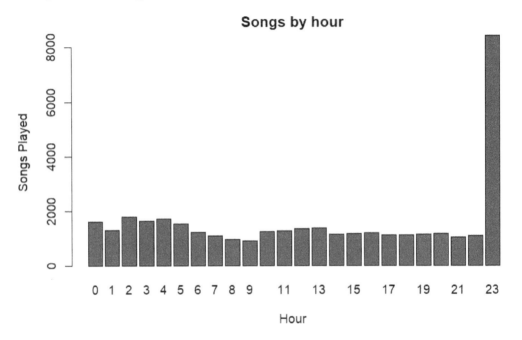

Figure 6.7 – Songs played by hour: sum of all the days

It looks like there is something going on at 11 pm. (23 hours). Let's filter the data by that time and take a closer look, just as we would do in a real-life project. It could be some duplicated values or just that 23 hours is when there are a lot of songs being played.

We can create a filtered dataset with songs that played at 23 hours using the next code:

```
# Filter only songs played at 23h
df_23 <- df[df$hour == 23,]
```

Here, we can use a little help from the tidyverse package, with the arrange function, to order the dataset after filtering:

```
df_23 <- arrange(df_23, `Song Clean`,CALLSIGN, TIME)
```

The resultant data frame is shown in *Figure 6.8*:

	Song Clean	ARTIST CLEAN	CALLSIGN	TIME
1	(Don't Fear) The Reaper	Blue Oyster Cult	KZEP	2014-06-18 23:58:51
2	(Don't Fear) The Reaper	Blue Oyster Cult	KZOK	2014-06-17 23:54:27
3	(Don't Fear) The Reaper	Blue Oyster Cult	KZOK	2014-06-18 23:54:22
4	(Don't Fear) The Reaper	Blue Oyster Cult	KZOK	2014-06-20 23:54:26
5	(Don't Fear) The Reaper	Blue Oyster Cult	KZOK	2014-06-21 23:54:23
6	(Don't Fear) The Reaper	Blue Oyster Cult	KZPS	2014-06-17 23:58:04
7	(Don't Fear) The Reaper	Blue Oyster Cult	WCSX	2014-06-16 23:54:15
8	(Don't Fear) The Reaper	Blue Oyster Cult	WCSX	2014-06-17 23:54:06
9	(Don't Fear) The Reaper	Blue Oyster Cult	WCSX	2014-06-18 23:54:03
10	(Don't Fear) The Reaper	Blue Oyster Cult	WCSX	2014-06-19 23:54:25
11	(Don't Fear) The Reaper	Blue Oyster Cult	WCSX	2014-06-20 23:54:07
12	(Don't Fear) The Reaper	Blue Oyster Cult	WCSX	2014-06-21 23:54:03

Figure 6.8 – Songs played at 23 hours

Figure 6.8 shows a partial screenshot of the sliced dataset, where we observed that either a song is being played by the same radio, but on different days, or it is being played by different radio stations on the same day. It looks like a lot of duplicates, but we can make sure that they are not by performing a test. For that, we will go once more for the `tidyverse` functions and we will add a different component—that is, the `%>%` pipe symbol, utilized to connect functions. All it does is say: *Take the result of the function and use it as input for the next function.*

Let's do the test. Take the number of rows in the dataset and subtract the number of unique observations:

```
# Checking if there are duplicated rows
# Number of rows in the dataset - Number of unique rows
dim(df_23)[1] - df_23 %>% distinct() %>% nrow()
[1] 280
```

The result is 280 observations. So, let's look at them a little closer by isolating those rows in a separate dataset called dups:

```
# Filter only the duplicated rows
dups <- df_23[duplicated(df_23),]
# Dimensions of dups
dim(dups)
[1] 280   11
```

Now, a `real_dup` empty vector is created. Then, there is a `for` loop for each row, comparing the current row (dups[row,]) with the subsequent one (dups[row+1,]). Keep in mind that dups has 11 columns and that R considers Boolean values such as TRUE equal to 1 and FALSE equal to 0. So, if the comparison between both rows is all TRUE values, it should add up to 11. If any variable is different, then it will be different than 11. This is enough for our test. We ask that only those rows where the comparisons return 11 are to be added to the `real_dup` vector. The code can be seen in the following sequence:

```
# Test for real duplicates: if sum of TRUE equal cells is 11
(all columns) then it is dup.
real_dup <- c()
for (row in 1:279) {
  equal_cols = sum(dups[row,] == dups[row+1,])
    if (equal_cols == 11) {real_dup <- c(real_dup, row) } }
```

If we call `real_dup` now, it returns rows 9 and 81, confirming that there are only two real duplicated observations in our data, making it easier to check:

```
real_dup
[1]   9 81
# Look at duplicates
dups[c(9,81),]
```

These are the two duplicated rows:

Song Clean <chr>	ARTIST CLEAN <chr>	CALLSIGN <chr>	TIME ▶ <S3: POSIXct>
Another One Bites the Dust	Queen	WCSX	2014-06-19 23:54:25
Fly Like an Eagle	Steve Miller Band	WCSX	2014-06-19 23:54:25

2 rows | 1-4 of 11 columns

Figure 6.9 – Duplicated rows

We still don't understand why these songs are duplicated. So, we can go back to the original dataset to filter it using the information from the duplicated observations, using the TIME and COMBINED (combination of song and artist), variables, using the following code:

```
# Checking duplicates
df_original %>%
  filter( (TIME == ymd_hms("2014-06-19 23:54:25")) &
          (COMBINED %in% c('Another One Bites the Dust by
Queen',
```

```
                        'Fly Like an Eagle by Steve Miller
Band')))
```

We will see this result:

ARTIST CLEAN	CALLSIGN	TIME	UNIQUE_ID	COMBINED
<chr>	<chr>	<dbl>	<chr>	<chr>
Queen	WCSX	1403222065	WCSX0834	Another One Bites the Dust by Queen
Queen	WCSX	1403222065	WCSX0949	Another One Bites the Dust by Queen
Queen	WCSX	1403222065	WCSX1021	Another One Bites the Dust by Queen
Steve Miller Band	WCSX	1403222065	WCSX0813	Fly Like an Eagle by Steve Miller Band
Steve Miller Band	WCSX	1403222065	WCSX0955	Fly Like an Eagle by Steve Miller Band
Steve Miller Band	WCSX	1403222065	WCSX1005	Fly Like an Eagle by Steve Miller Band

Figure 6.10 – Duplicated rows in the original dataset

We found the problem. The observations have different UNIQUE_ID values. They appear three times each, actually. But as they have different IDs, it is up to us, data scientists, to remove them or not. Since that is only four entries we can remove them, recreate the df data frame following the same code previously written, and move forward with the analysis. You can see the cleanup code in GitHub (https://tinyurl.com/khfrrnae).

Another check we can do is to plot the average of songs played by hour by weekday. The previous plot from *Figure 6.7* shows the sum of the songs played by hour for all the days in that week. Maybe a better approach would be to group the dataset by day and take an average of songs played by hour.

That is performed in the following sequence, where we are using the tidyverse package to perform this task. The first code snippet is taking the df dataset, grouping the data by weekday and hour, and counting the number of songs by group. The result would be a data frame of songs played by hour and by each day in a week:

```
# Group data by weekday
df_by_day <- df %>%   #take the df object
  group_by(weekday, hour) %>%   #group by weekday and hour
  summarise(songs_ct = n() ) %>%   #count songs per group
```

But we still want to take an average of songs by hour, considering all days of the week. The result of the last piece of code will be one number for each weekday and hour (at noon, for example, 181 on Monday, 212 on Tuesday, 157 on Wednesday, up until Sunday). We want to have a single average number for noon. Thus, we can connect the interim result with another group_by() function by hour and calculate the mean of how many songs are played by hour:

```
# Group the result by hour
df_by_day <- df_by_day %>%
```

```
group_by(hour) %>%  # group the result by hour
summarize(avg_songs_ct = mean(songs_ct)) #count the avg of
songs
```

Next, we plot the result in a bar plot:

```
# Line plot of Songs by weekday
plot(y= df_by_day$avg_songs_ct, df_by_day$hour,
     type = "l", pch = 19, lwd=3,
     col="royalblue",
     main="Avg Number of Songs Played by hour",
     xlab="Hour",
     ylab= "Songs Played")
```

There is a difference between both plots. The first one was a bar plot. The next is a line plot. The distribution is similar, but note that the *y* axis in *Figure 6.11* is much smaller now, down to 1,200 against over 8,000 from *Figure 6.7*. Plotting a bar plot and a line plot to compare them is not the best option, for sure, but I did that just to show how lines are a better fit for plots based on time because it is easier to spot peaks and valleys. In *Figure 6.11*, we have a complete view of an average day, by hour:

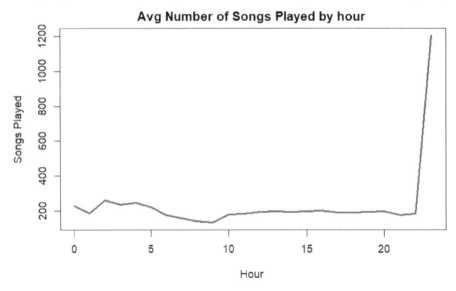

Figure 6.11 – Line plot of average number of songs by hour

The spike at 23 hours is real. We saw that there were no duplicates. So, it possibly reflects a radio station's schedule pattern. Maybe at that time every day, there's a commercial-free show.

Continuing this practice, let's see *which hour has the highest number of songs being played for the first time.*

We create a filtered dataset with songs played for the first time only:

```
# Filter only first time = 1
first_time <- df[df$`First?`==1,]
```

We group by hour and count the number of songs:

```
# First Time Songs by Hour
first_time <- first_time %>%
  group_by(hour) %>%
  summarise(song_ct= n())
```

Next, the code for the line plot:

```
# Line plot of First appearance Songs by hour
plot(x=first_time$hour, y=first_time$song_ct,
     col="royalblue", type="l", lwd=3,
       main="First Time Played by hour",
       xlab="Hour", ylab= "Songs Played")
```

And the result is next:

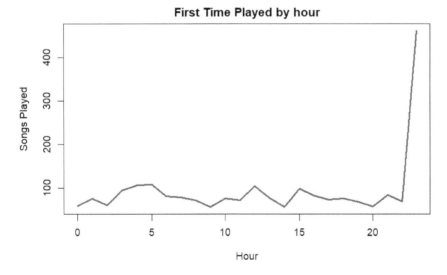

Figure 6.12 – Line plot of average number of first-time songs by hour

Interestingly, the distribution in *Figure 6.12* also follows the distribution of the average of songs played by hour (*Figure 6.11*). Besides 11:00 pm, there are other peaks around 5 am, noon, 3 pm, and 9 pm Those are the best times to discover new songs, according to our data.

Finally, now we will plot a bar plot with the top five radio stations that played more songs in that week of the dataset. Once again, we start creating a grouped data frame using the following code. Check the comments to see what each line is doing. Once again, don't worry, as we will study all of these functions from `tidyverse` in *Chapter 8*:

```
# Average by radio by day
by_radio <- df %>%  # take the dataframe
  group_by(CALLSIGN, weekday) %>%  #group by Call sign and weekday
  summarise(song_ct = n()) %>%  #Count the observations by group
  group_by(CALLSIGN) %>%  #group again only by call sign
  summarise(avg_song= mean(song_ct)) %>%  # calculate average of songs
  arrange(desc(avg_song)) %>%  # organize in descending order
  head(5) # keep only the first 5 rows of the result
```

And to plot the result, the code is shown next:

```
# Bar Plot
barplot(avg_song ~ CALLSIGN, data=by_radio,
        horiz = T,
        main='Average Number of Songs by Radio Station in a Day',
        xlab = 'Avg number of songs played', ylab='Radio Station',
        col='royalblue')
```

Figure 6.13 leads us to conclude that *KSEG* can brag about being the radio station that plays more new songs:

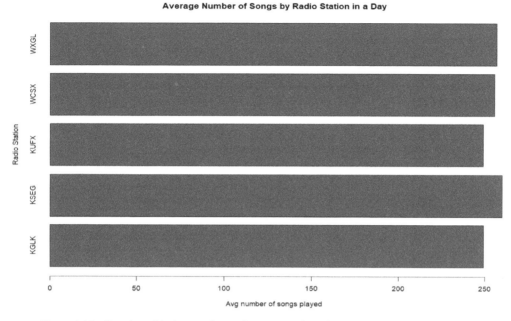

Figure 6.13 – Bar plot with the top five radio stations that play more songs for the first time

With that, we end this practical exercise, where the goal was to show how to devise questions to help you to explore the data. There is much more that could be explored and done with this data, but I believe that what was covered here is good enough to understand the usage of datetime variables in practice.

Summary

We progressed a lot in this chapter and learned so much about datetime objects and variables. Knowing how to use them will enhance your analytical skills, opening room for better insights.

We started this chapter by learning how to create data objects. Next, we acquired knowledge on how to make good use of the `lubridate` library, and we are now able to parse dates in many different formats.

After that, the subject changed to math operations with date and time and all the specificities that surround it, including the usage of time zones. That was followed by guidance on how to use the customization power of `regexp` to parse dates out of texts.

Closing the chapter, we viewed a practical exercise where we used a dataset about songs containing a datetime variable, and how you can use that to extract insight for analysis.

The end of this chapter is also the completion of the building blocks for data wrangling. With it, we now have worked with the three major object types in R: *strings*, *numbers*, and *datetime* objects. In the next chapters, it is time to start transforming data, which I consider a central part of the data-wrangling process.

Exercises

1. What are date, time, and datetime objects?

2. Name one function used to create a datetime object.

3. List some of the parsing functions from `lubridate` to extract periods of time from datetime objects.

4. What is a period object?

5. What is an example of usage for the duration object?

6. How can you add or subtract dates?

Further reading

- Explore more about `lubridate` in the official documentation: `https://lubridate.tidyverse.org/`

- Learn more about `tidyverse`: `https://www.tidyverse.org/`

- Here's the dataset used in the practical exercise in this chapter:

    `https://github.com/fivethirtyeight/data/tree/master/classic-rock`

- Here's the code for this chapter on GitHub: `https://tinyurl.com/yckh7w4u`

7
Transformations with Base R

In the last three chapters of this book, our intent was to lay the foundations of the main data types you will find when working on a real-life project. Once a dataset is opened, it is likely you will find strings, numbers, and dates and times as variables. Knowing what they are, how they can be created, and some popular functions to manipulate them will keep us moving during exploratory data analysis.

The next two chapters are focused on transformations of data, what I consider the core of data wrangling. I say that because the main part of what needs to be done during the data wrangling phase of a project will be related to transformations of data. Then, another good chunk of the work is on data visualization, and the final piece will be modeling and evaluating the results.

In this chapter, we will study the most common transformations that can be done in a dataset:

- **Slicing** and **filtering**: These tasks allow us to focus on a specific part of the dataset.
- **Grouping** and **summarizing**: These are important functions that reduce the number of rows in a dataset, providing a single number that summarizes a measurement about a group.
- **Replacing** and **filling**: These functions fill gaps and replace values with new ones.
- **Arranging**: We can ordinate the data to create a rank or have an idea of the proportion of the observations.
- **Creating new variables**: Data wrangling is not only about manipulating existent data, but also about creating new variables as combinations of others.
- **Merging**: Datasets can be complementary, so a variable from B may be needed in A. The solution is to merge the datasets.

We will start with transformations using R's built-in methods again because they are easy to use and do not require any package installations, thus we will get no dependency errors. Another resource to be used is the data.table library. Since base R does not have all the methods we need, this library will cover those gaps.

The theory presented in this chapter will be valid for the next chapter too, and you will be learning two ways to wrangle data, enlarging your skillset. However, I realize that you might prefer the transformations done with the tidyverse packages and might use them more regularly in your work.

Technical requirements

We will use the Census Income dataset (https://archive.ics.uci.edu/ml/datasets/ Adult) for this chapter.

All the code can be found in the book's GitHub repository: https://github.com/ PacktPublishing/Data-Wrangling-with-R/tree/main/Part2/Chapter7.

Before moving forward, make sure to run the following installation requirements if you want to code along with the book's examples:

```
# Install package
install.packages('data.table')
# Load library
library(data.table)
library(stringr)
```

The dataset

The dataset to be used in the next exercises can be found in the UCI dataset repository. It was pulled from the *popular datasets* tab in the repository, and it is named Adults, but it is also known as *Census Income* dataset (from https://archive.ics.uci.edu/ml/datasets/Adult).

The variables we will be dealing with are listed next and I also invite you to read the adult. names file provided with the dataset in the UCI repository or in the GitHub page for this chapter's codes (https://tinyurl.com/ywpjj329).

- Demographics: *age, sex, race, marital-status, relationship status, native country.*
- Education: *education* level and *education-num* (years of study).
- Work related: *work class, occupation, hours per week.*
- Financial: *capital gain* and *capital loss.*
- *fnlwgt*: This means final weight, which is a scoring calculation from the Census Bureau based on socio-economic and demographic data. People with similar demographic information should have similar weights.

One final note about the dataset is that NA values are converted to ?.

The dataset can be loaded directly from the web into RStudio using the following code snippet:

```
# Load the dataset to Rstudio
url <- 'https://archive.ics.uci.edu/ml/machine-learning-
databases/adult/adult.data'
df <- read.csv(url, header=FALSE, strip.white = TRUE)
```

We must start our data wrangling early with this dataset. Here, we added `strip.white=TRUE` to solve the problem of extra blank spaces before string values in the variables, as the parsing was not perfect after collecting the dataset from the web. Another addition is the `header=FALSE` parameter because the dataset is provided as one file for the data and the variable names are described on the UCI web page. So, we will add a vector with the variable names to the `colnames()` function. When you do that, the vector will overwrite the column names with the vector presented by you.

```
# Add column names
colnames(df) <- c('age', 'workclass', 'fnlwgt','education',
'education_num', 'marital_status', 'occupation',
'relationship', 'race', 'sex', 'capital_gain', 'capital_
loss','hours_per_week', 'native_country', 'target')
```

You can use `head(df)` or `View(df)` to see the dataset and check whether the columns have been properly placed.

	age <int>	workclass <chr>	fnlwgt <int>	education <chr>	education_num <int>	marital_status <chr>
1	39	State-gov	77516	Bachelors	13	Never-married
2	50	Self-emp-not-inc	83311	Bachelors	13	Married-civ-spouse
3	38	Private	215646	HS-grad	9	Divorced
4	53	Private	234721	11th	7	Married-civ-spouse
5	28	Private	338409	Bachelors	13	Married-civ-spouse
6	37	Private	284582	Masters	14	Married-civ-spouse

Figure 7.1 – Dataset loaded and variable names added

The dataset is now ready to be wrangled; we will do that in the next sections.

Slicing and filtering

When you have a table as large as the dataset we are working with, it is very hard to look at all the observations one by one. Look how many rows and columns this dataset has:

```
# Dataset dimensions
dim(df)
[1] 32561     15
```

The dim() function shows the number of rows first, then the number of columns, or variables. It's easy to see that it would take us too much time – not to mention that it is not productive as well – to look at 32,561 observations. Therefore, the tasks of slicing and filtering play a major role, acting like a magnifying glass for us to zoom in on specific parts of the data.

These tasks can sound like they're the same, but there is a slight difference between them.

Slicing

Slicing means cutting and displaying a slice, a piece, of the dataset. A good application of this task is when we need to look at the errors of a model. In this case, it is possible to take only the observations where we see classification errors, for example, and analyze them separately. Another similar example is to remove outliers from a dataset as a cleaning task during data wrangling.

Slices can be made only for rows, only for columns, or both, and the syntax is composed of the object followed by square brackets, and inside those brackets we add the row number before the comma and the column number after the comma. It accepts conditional tests replacing rows or columns, and leaving a blank space means return all the rows or columns.

Next, we present a way to slice rows by their number and fetch all columns. There is a blank space after the comma, denoting no slicing for the columns; therefore, this slice returns all of the columns:

```
# Slicing rows 1 to 3, all columns
df[1:3,]
```

The result is shown in *Figure 7.2*.

	age <int>	workclass <chr>	fnlwgt <int>	education <chr>	education_num <int>	marital_status <chr>
1	39	State-gov	77516	Bachelors	13	Never-married
2	50	Self-emp-not-inc	83311	Bachelors	13	Married-civ-spouse
3	38	Private	215646	HS-grad	9	Divorced

3 rows | 1-7 of 15 columns

Figure 7.2 – A slice of only the first three observations

The following code snippet shows how to slice data by columns but fetch all the rows. Notice that the blank space is before the comma in this case:

```
# Slicing all rows, columns 1 to 3
df[,1:3]
```

This returns the table shown in *Figure 7.3*.

age <int>	workclass <chr>	fnlwgt <int>
39	State-gov	77516
50	Self-emp-not-inc	83311
38	Private	215646
53	Private	234721
28	Private	338409
37	Private	284582
49	Private	160187
52	Self-emp-not-inc	209642
31	Private	45781
42	Private	159449

1-10 of 32,561 rows

Figure 7.3 – A slice of only the first three columns from left to right

The next code is used to slice the data based on a condition, using a vector with the numbers of the columns to display:

```
# Slicing with conditional, vector for cols 1 and 6
df[df$age > 30 ,c(1,6)]
```

We are asking R to return people who are over 30 years old, and only columns number 1 (age) and 6 (marital_status). The result is displayed in *Figure 7.4*.

	age <int>	marital_status <chr>
1	39	Never-married
2	50	Married-civ-spouse
3	38	Divorced
4	53	Married-civ-spouse
6	37	Married-civ-spouse
7	49	Married-spouse-absent
8	52	Married-civ-spouse
9	31	Never-married
10	42	Married-civ-spouse
11	37	Married-civ-spouse

1-10 of 21,989 rows

Figure 7.4 – A slice of people over 30 years old and columns 1 and 6

Slicing is a simple, yet powerful task, especially if you are working with big data. When you have enormous amounts of data, inputting all of it into a function or model can overwhelm the system, and consequently, your script may underperform. Slicing or filtering it decreases the size of the dataset, making it easier for the computer to handle it in memory.

Filtering is a lot like slicing, but it's not the same. Let's see why next.

Filtering

The difference between slicing and filtering is that, when filtering data, the return will be all the variables. If we filter to return only people over 30 years old, the result will be all the observations with all the variables that fulfill that condition. Observe the following code, where the `subset()` function is used. It receives a dataset and a condition as input:

```
# Filter
subset(df, age > 30)
```

	age <int>	workclass <chr>	fnlwgt <int>	education <chr>	education_num <int>	marital_status <chr>
1	39	State-gov	77516	Bachelors	13	Never-married
2	50	Self-emp-not-inc	83311	Bachelors	13	Married-civ-spouse
3	38	Private	215646	HS-grad	9	Divorced
4	53	Private	234721	11th	7	Married-civ-spouse
6	37	Private	284582	Masters	14	Married-civ-spouse
7	49	Private	160187	9th	5	Married-spouse-absent
8	52	Self-emp-not-inc	209642	HS-grad	9	Married-civ-spouse
9	31	Private	45781	Masters	14	Never-married
10	42	Private	159449	Bachelors	13	Married-civ-spouse
11	37	Private	280464	Some-college	10	Married-civ-spouse

1-10 of 21,989 rows | 1-7 of 15 columns   1   2   3   4   5   6 _ 100

Figure 7.5 – Filtered dataset: observations where age is over 30

The `subset` function can also help you to slice data. If you add the `select` parameter, it is possible to display the result with just the columns that we want:

```
# Slice with subset, returning only the selected columns
subset( df, age > 30, select=c(age, marital_status) )
```

The result of this code is the same as *Figure 7.4*.

This concludes the concepts of slicing and filtering. We have used them before and will keep using them many other times in this book. Next, we will introduce the grouping and summarizing functions, a duo that does wonders for data analysis and data wrangling.

Grouping and summarizing

The same logic used to present the slicing and filtering concepts can be applied here too: we will never go row by row, analyzing one observation at a time.

We need a better way to look at the data, one that makes it smaller and easier to understand. To do that, we can aggregate data, creating groups of observations and putting each one of them in a separate and labeled box. This is grouping.

After that, we have groups, but we still don't have a very good use for *n* boxes that we don't know the contents of, besides the name of the group on the *label*. Summarization will do that job by taking the observations in each box and wrapping them up with a single number, which could be the mean, the median, or the total. Summarization is, therefore, reducing observations to one number.

Given these definitions, it is reasonable to say that summarization is complementary to the grouping function since we first aggregate the data in groups and then summarize what is happening in that group with a measure.

Performing this task in base R is tricky, so we will use the `data.table` library for this task. First, we need to convert the data frame object to a `data.table` object and then group it.

If we want to know if there are more men or women in our dataset, we can group the observations by sex and count them. The task is performed by the next lines of code:

```
# Data frame to Data.table
dt <- as.data.table(df)


# Group By and count number of observations
dt[, .N, by = sex]
```

The grouping with data.table is done using slicing notation. You provide the dataset as a `data.table` object, and all the rows will be considered for the calculation since the space before the comma is empty. `.N` is a special variable that counts the number of rows in each group, and `by = sex` gives us the variable we want to group by. So, it can be read as *from the `dt` object, count the observations by each group from the sex variable (Male and Female)*. As a result, we see the number of observations for men and women in *Figure 7.6*. There are approximately twice as many men as women in the dataset.

sex <chr>	N <int>
Male	21790
Female	10771

2 rows

Figure 7.6 – Count of observations by sex

Alternatively, you can pass another function, such as mean(), median(), or sum(), as the second argument. For example, what is the average age of men and women in the data? Let's find out:

```
# Group By and return mean of age by sex
dt[, mean(age), by = sex]
```

The result is presented in *Figure 7.7*, where we see that the men in the dataset are, on average, three years older than the women.

sex <chr>	V1 <dbl>
Male	39.43355
Female	36.85823

2 rows

Figure 7.7 – Average age by sex

Grouping and summarizing are crucial for analyzing and wrangling data. Many times, the table resulting from a grouping task can become the input for a visualization. Let's see how.

The data has a variable that shows the number of years a person was educated and the level of education achieved. If we want to know how many years, on average, it takes for a person to achieve that education level based on our data, we should start grouping the observations by education level and then take the average number of years in school for each education level. The result will be saved in educ_yrs. That is what the next code does:

```
# Mean education years by education level
educ_yrs <- dt[, mean(education_num), by= education]
educ_yrs <- educ_yrs[order(V1)]
```

The result, educ_yrs, is used to create a bar plot, with the following code:

```
# Bar plot
barplot(educ_yrs$V1, names.arg = educ_yrs$education,
col="royalblue",
        main="Average years of education by grade level")
```

Figure 7.8 shows the bar plot.

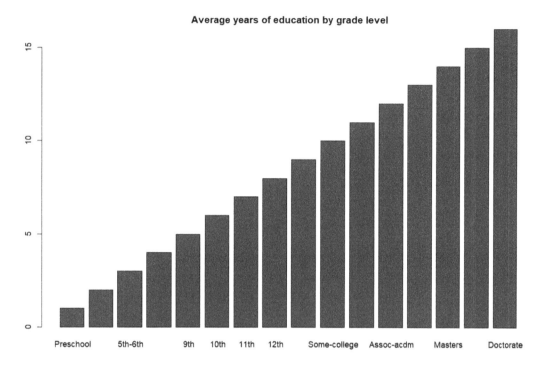

Figure 7.8 – A graphic created using the output data from a grouping function

The result reflects the reality, knowing that it takes more years in school to reach higher education levels. Now that we know more about grouping and summarizing, let's move on to learn how to replace and fill in missing data.

Replacing and filling

Replacing values is straightforward. You have a value that does not fit in the data and you need it to be replaced with another value. In the dataset we're using in this chapter, there is a good example. In the documentation about the data, it is stated that the author will *convert unknown values to "?"*, meaning that you will not find any standard NA values in this dataset. Therefore, it is our job as data scientists to wrangle this and replace all the ? values with NA.

> **Note**
>
> It's worth making a note of this, as a lesson learned from this exercise: always look at the data documentation, if and when it is available. Many explanations about the way the data was collected and the meaning of each variable are contained in these documents.

Replacing the values is possible using slicing notation or the gsub() function. In the dataset, there are three variables with ? values: workclass, occupation, and native_country.

We will replace them with the following code, using slicing and using gsub(). The first option, using slicing notation is read as *take the occupation variable, slice only the* ? *values, and assign* NA *to them.* The second should be read as *from the dataset, slice the rows where* native_country *is equal to* ? *and the* native_country *column and assign* NA *to those values*:

```
# Replace "?" using slicing
df$occupation[df$occupation == "?"] <- NA
df[df$native_country == "?", "native_country"] <- NA
```

The next one is *substitute* ? *values with* NA *values in the* workclass *variable*:

```
# Replace "?" with NA using gsub
df$workclass <- gsub(pattern= "\\?", replacement= NA,
df$workclass)
```

The replacement can also be done using logical tests, performed by the ifelse() function from base R. If the test is true, ? values are replaced with NA, otherwise it keeps the original value in the cell:

```
# Replace "?" with ifelse
df$native_country <- ifelse(df$native_country == "?", NA,
df$native_country)
```

These are the basics of replacing, but we've also covered most of what you will need in terms of replacements.

Now that we've got rid of the question marks, our dataset is full of missing values, the notorious NA. To make sure that the transformations worked, we can run a sum of NA values using a combination of sum() and is.na(), as shown in the following snippet:

```
# Check for NA values
sum(is.na(df))¶[1] 4262
```

There are over four thousand NA values now. In this case, it is necessary to fill in the missing values or remove them. As we mentioned in *Chapter 2*, if the missing data makes up at most 5% of the data, it is safe to remove it without much interference in our analysis.

To find out how many missing values we have by variable, the easiest way is to load the skimr library and run skim(df). This command will provide the result shown in *Figure 7.9*, where we can see that the workclass and occupation variables have a little more than 5% of missing values each. Remembering the dimensions we read at the beginning of this chapter, this dataset has 32,561 rows.

For `workclass`, the calculation is 1,836/32,561=0.056, or 5.6%. The result for `occupation` is very similar, also approximately 5.6%.

	skim_variable <chr>	n_missing <int>	complete_rate <dbl>	min <int>	max <int>	empty <int>	n_unique <int>
1	workclass	1836	0.9436135	8	17	0	8
2	education	0	1.0000000	4	13	0	16
3	marital_status	0	1.0000000	8	22	0	7
4	occupation	1843	0.9433985	6	18	0	14
5	relationship	0	1.0000000	5	15	0	6
6	race	0	1.0000000	6	19	0	5
7	sex	0	1.0000000	5	7	0	2
8	native_country	583	0.9820951	5	27	0	41
9	target	0	1.0000000	5	6	0	2

9 rows

Figure 7.9 – skimr is an easy way to count missing values by variable

We will replace the missing points with the most frequent value for these categorical variables. As the data wrangling gets more complicated, more difficult coding is needed from base R. Notice in the next code that, to replace the NA values with the most frequent value, we have to combine three functions: the `table()` function counts the occurrences by category, then `which.max()` gets the position of the element with maximum value, then `names()` is used to collect the name of that category. The result is stored in `most_frequent`:

```
#Fill NAs with most frequent value
most_frequent <- names(table(df$workclass)[which.
max(table(df$workclass))])
```

Next, we slice the dataset to get the observations where the values of the `workclass` variable are NA and assign the `most_frequent` value to them:

```
df$workclass[is.na(df$workclass)] <- most_frequent
```

This code is repeated for the `occupation` variable:

```
# Occupation
most_frequent <- names(table(df$occupation)[which.
max(table(df$occupation))])
df$occupation[is.na(df$occupation)] <- most_frequent
```

After that, there will only be the NA values for `native_country`, representing 1.7% of the total observations. Following the 5% rule, it is safe to remove all of those 583 NA observations from the dataset.

The function to use is `na.omit(df)`. This command will drop the observations with missing values from all the variables within `df`, so don't use this function if you haven't checked the missing values to see whether it is OK to drop them:

```
# Remove all NAs (now only the native_country are left)
df <- na.omit(df)
```

If we are using numbers, the function from base R for replacing NA values is `nafill()`. You can choose between filling with a constant, such as 0, filling with the last valid value (the previous row), or the next valid value.

Let's put our *Census Income* dataset aside for a minute and create a new dataset with numeric variables and some NA values for the next example. The code to create it is as follows:

```
# Data frame
df_num <- data.frame(A= c(1,2,2,2,3,NA),
                     B= c(3,4,5,3,NA,0),
                     C= c(1,1,1,NA,NA,5))
```

Now, to fill the NA values with zeros, here is the code:

```
# Fill NAs with 0
nafill(df_num, type='const', fill=0)
```

To fill with the last valid number or the next valid number, the `type` argument changes to `locf` or `nocb`:

```
# Fill NAs with the last valid number
nafill(df_num, type="locf")
```

```
# Fill NAs with the next valid number
nafill(df_num, type="nocb")
```

To fill the NA values with the mean, the code changes a little. We slice the data to get the NA values from column A and assign the calculated average of column A to replace them. The `na.rm=T` argument means that we are removing any NA values from the average calculation. See the following code:

```
# Fill NAs with mean
df_num$A[is.na(df_num$A)] <- mean(df_num$A, na.rm = T)
```

You can see the result of this last code snippet in *Figure 7.10*. Column A had an NA value in row 6, which was replaced with the mean value: 2.

▲	A	⇕	B	⇕	C	⇕
1	1		3		1	
2	2		4		1	
3	2		5		1	
4	2		3		NA	
5	3		NA		NA	
6	2		0		5	

Figure 7.10 – Row 6 on column A was filled with the mean value (2)

To continue our studies about transformations, we will see how to arrange data, a useful task for showing the top- or bottom-ranked items.

Arranging

Data can be arranged in two common forms: low to high or high to low, also known as ascending and descending order. Arranging data is useful for ranking observations or groups in an order that makes it easier for us to understand. When I look at the *top 5* most sold items, I know that they are what brings traffic to a store. Then, imagine that the third best-selling item in terms of count is, in fact, the product that makes the most revenue. That could change our strategy, couldn't it?

When looking at the other side of the rank, the tail, the bottom 5 items in terms of the number of items sold could be potential candidates to remove from the shelves, as they probably won't bring much revenue to the business.

This simple example explains why arranging data is important when exploring data. But arranging is also an important part of data wrangling when visualizing data because it quickly pulls our eyes to the maximum point, from where we can read the rest of the graphic.

Let's see how to arrange data with base R now. We can arrange the dataset with a single column. There are a couple of ways to do that. One is using the order() function with the decreasing=T parameter to sort values from high to low, or decreasing=F to sort from low to high:

```
# Arrange raw data
df[order(df$age, decreasing = T),]
```

The other is using a minus sign in front of the variable if we want to arrange it in descending order:

```
# Arrange raw data using - instead of "decreasing=T"
# remove the - for increasing order.
df[order(-df$age),]
```

As we've said, ordering data after grouping is a very effective way to understand the data quickly. The next code uses the data.table library. We are back with the *Census Income* dataset, and it has been transformed into a `data.table` object. To group it, we consider all the rows (using an empty space before the comma), calculate the mean of `education_num`, and group it by `workclass`. Then, we add some slicing notation `[order(-V1)]` to say that we want to arrange the result from high to low. Notice that we use `V1` because, after calculating the mean, the library changes the name of the variable to `V1`:

```
# Group and order education years average by workclass
dt[, mean(education_num), by= workclass][order(-V1),]
```

The result is presented next.

workclass <chr>	V1 <dbl>	workclass <chr>	V1 <dbl>
State-gov	11.349492	State-gov	11.349492
Self-emp-not-inc	10.211285	Self-emp-inc	11.167598
Private	9.824196	Local-gov	11.036768
Federal-gov	10.948038	Federal-gov	10.948038
Local-gov	11.036768	Self-emp-not-inc	10.211285
Self-emp-inc	11.167598	Private	9.824196
Without-pay	9.071429	Without-pay	9.071429
Never-worked	7.428571	Never-worked	7.428571

Figure 7.11 – Comparison of data before and after being arranged

Figure 7.11 compares the unordered result with the ordered result to illustrate how much easier it is to identify the top three work classes in terms of their mean years of study. In the table, there are many values that are very close, making it hard for us to quickly determine the top ones. In the right-hand table, this is easier. People who work for the US government at many levels, along with self-employed people, are those with more years of study, on average, for this dataset.

Next, let's see how to create new variables in R, a task commonly performed when wrangling data.

Creating new variables

A dataset is not only the data you see. There is a lot of information in it. For example, remember in *Chapter 6*, when we worked with datetime objects during our data exploration exercise, we took the `TIME` variable and extracted the year, month, day, and hour from it. This is one of the many ways to create new variables.

Here are some examples of new variables created out of our working dataset:

- **Arithmetical operators**: Adding two or more variables to create a `total` variable.

- **Text extraction**: Extracting a meaningful part of a text, for instance, `1234` from `ORDER-1234`.

- **Custom calculations**: Calculating a discount rate based on a business rule.

- **Binarization**: Transforming a variable from `on` and `off` to `1` and `0`. Binary means two options and is commonly associated with `0` and `1` in computer language.

- **Encoding**: Transforming a qualitative ordinal variable, such as `basic`, `intermediate`, and `advanced` to `1`, `2`, and `3`.

- **One Hot Encoding**: A very common transformation for machine learning, it means to transform a qualitative variable with many levels into columns with 0s and 1s, where a 1 will be the category for that observation. For example: if the groups are A, B, and C, there would be three new columns, and an observation from group A has the values 1, 0, 0 for the three columns.

Group	A	B	C
A	1	0	0
B	0	1	0
C	0	0	1

Figure 7.12 – An example of one hot encoding with three groups

To create new columns in R, there are two possible codes, presented here. Personally, I think it is easier to go with the first code, using the dollar sign, $, but you do it whichever way feels more comfortable. Here are the codes for the two options:

```
df$new_col_name <- values
df['new_col_name] <- values
```

With that in mind, we can create a binary version of the `sex` variable:

```
df$sex_binary <- ifelse(df$sex == 'Female', 1, 0)
```

	sex <chr>	sex_binary <dbl>
1	Male	0
2	Male	0
3	Male	0
4	Male	0
5	Female	1
6	Female	1
7	Female	1

Figure 7.13 – Example of a binarized sex variable

Encoding a variable transforms the text into numbers, making the data more suitable for many machine learning algorithms. In the following code, we will transform the education variable into a coded version. First, we must convert the variable from character to factor, which is the same as categories in R.

```
# Assign education as factor
df$education <- as.factor(df$education)
```

Then, to create an order of the factors, we assign an ordered vector to the education variable:

```
# Order the factors
df$education <- ordered(df$education, levels= c("Preschool",
"1st-4th","5th-6th","7th-8th", "9th", "10th", "11th",
"12th", "HS-grad", "Some-college", "Assoc-acdm", "Assoc-voc",
"Bachelors", "Masters" , "Doctorate") )
```

Next, we assign the ordered education variable to a new variable named education_cd and change the levels to numbers:

```
# Create new variable
df$education_cd <- df$education
# Change levels to numbers
levels(df$education_cd) <- 1:15
```

The original and the new variable are printed in *Figure 7.14*.

	education <ord>	education_cd <ord>
1	Bachelors	13
2	Bachelors	13
3	HS-grad	9
4	11th	7
5	Bachelors	13
6	Masters	14
7	9th	5

Figure 7.14 – The education variable was encoded and a new variable was created

The new column, `education_cd`, has numbers for each of the grade levels now.

Arithmetic operations with columns can be performed easily by choosing the variables, the math operation desired, and assigning the result to a new variable. In the next example, we'll see how to do that by creating a variable using custom calculations. Imagine that there is a business rule for our Census data saying that, for every `total_gain` equal to or over 15,000, there is a tax of 10%. One possible solution is to create a new variable named `total_gain` that is the result of the `capital_gain` – `capital_loss` calculation, as the code shows:

```
# Total gain variable creation
df$total_gain <- df$capital_gain - df$capital_loss
```

We can apply the tax business rule to `total_gain`. That implies the creation of a variable called `tax` for storing the calculation result. We will use the `ifelse()` function and test whether `total_gain` is over 15,000. If yes, then we apply the multiplication by 10%; otherwise, it is 0:

```
# Tax variable creation
df$tax <- ifelse(df$total_gain >= 15000, df$total_gain *0.1, 0)
```

You can see the result of the last calculations in *Figure 7.15*.

capital_gain <int>	capital_loss <int>	total_gain <int>	tax <dbl>
25236	0	25236	2523.6
25236	0	25236	2523.6
25236	0	25236	2523.6
25124	0	25124	2512.4
25124	0	25124	2512.4
25124	0	25124	2512.4
25124	0	25124	2512.4

Figure 7.15 – Two variables created using arithmetic operations

The `total_gain` variable may not even be needed. It is just a means to get to the tax amount calculation. We could have placed the subtraction operation directly in the `ifelse()` function, where it says `df$total_gain`.

Creating new variables is a valuable resource for data wrangling and uncovering good information hidden in the dataset. In the following section, we will cover the last main transformation for data wrangling: merging data. Frequently, data will be in pieces or will come from separate sources, thus we need functions to glue them together. That is coming up next.

Binding

Binding data is the last of the main transformations listed at the beginning of this chapter. It is common to find yourself with two or more datasets that you need to put together for analysis. There are a couple of ways to do that, as follows:

Task	Description	R function	Example
Append or bind rows	Dataset A and dataset B have the same variables, like two separate pieces of the same whole. It includes the *rows* from B at the bottom, after the last row of A.	`rbind()`	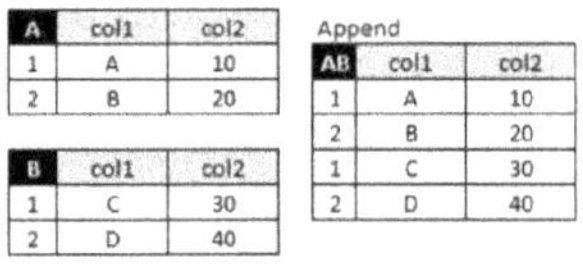 A: coll1=A col2=10 (1), B col2=20 (2); B: coll1=C col2=30 (1), D col2=40 (2). Append → AB: coll1 col2: 1 A 10, 2 B 20, 1 C 30, 2 D 40
Bind columns	Dataset A and dataset B have the same number of rows and index, like two sets of variables for the same observation. It includes the *columns* from B after the last column of A on the right-hand side.	`cbind()`	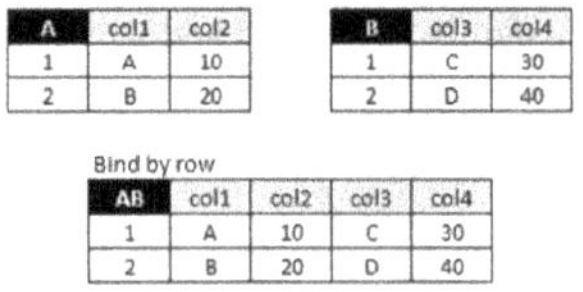 A: coll1=A col2=10 (1), B col2=20 (2); B: col3=C col4=30 (1), D col4=40 (2). Bind by row → AB: coll1 col2 col3 col4: 1 A 10 C 30, 2 B 20 D 40

Figure 7.16 – Types of data binding

Assume that our *Census Income* dataset has only 10 rows. After some research, the internal team found another 10 observations and gave them to the data science team. The ten new observations have to be appended to the original dataset since they have the same variables. Let's see that in action:

```
# Creating datasets A and B
A <- df[1:10, ]
```

```
B <- df[11:20, ]


# Append / bind rows
AB <- rbind(A, B)
```

To illustrate the other scenario, that is, binding columns, imagine that the original data has only three variables, age, workclass, and fnlwgt. Then, the team was able to collect more information about the taxpayers, adding education grade and occupation. The data now refers to the same observations, the same taxpayers, so we need to bind their columns using the following code:

```
# Creating datasets A and B
A <- df[1:10, 1:3]
B <- df[1:10, c(4,7)]


# Append / bind rows
AB <- cbind(A, B)
```

As seen, the code is practically the same, but it is necessary to assess the situation and determine what kind of binding to use. If the datasets have the same variables, then we are binding rows with rbind(). If the observations are the same but with different variables, we will bind columns with cbind(), inputting the datasets to be gathered to the functions.

Before we end the chapter, let's learn a little bit more about the data.table library and go over some of the ways it builds on the base R functions.

Using data.table

The data.table library describes itself as an *enhanced version of the data.frames* in R. Using only base R, it is not easy to group data, for example. There are other small enhancements, such as not converting strings to factors during data import and in the visualization of printing datasets on R's console.

The syntax for this library is very similar to data.frames, as you may have already seen during this chapter, but it is formally presented here:

```
Basic syntax
DT[i, j, by]
```

- i is for the row selection or a condition for the rows to be displayed

- j is for selecting variables or calculating a statistic based on them

- by is used for grouping variables

Before using the syntax for data.table, it is necessary to make sure that the object is the correct type. That can be done using `type(object)`. Conversion to a `data.table` object can be done using `as.data.table(object)`.

Consider the following code snippet:

```
# Syntax
dt[dt$age > 50, .(age, occupation)]
```

If you are familiar with SQL, you can read the `i` as a `WHERE` statement, `j` as a `SELECT` statement, and `by` as `GROUP BY`. Therefore, the code just presented is read as `SELECT age, occupation from dt WHERE age > 50`.

While `data.frame` uses a concatenated vector, `c(1,2)`, with the numbers or names of the variables desired for the slice, in data.table the syntax is slightly different: it uses `.(c1,c2)`.

To group by a given variable, as we have seen before, the code should use by within the slicing notation. Remember that by can receive more than one variable:

```
# Mean age Group By relationship
dt[, mean(age), by= relationship]
```

The blank space before the first comma means that all the rows will be considered for the grouping and mean calculation. If an ordered result is required, just add `[order(-V1),]` in front of the closing square bracket.

Base R and data.table are basic libraries, and they lack more complex functions. Therefore, when the problems become more complex, you may have noticed during this chapter that the number of lines of code increase proportionally, requiring us to create more interim variables or use other functions to solve them. In the next chapter, these same transformations will be solved with tidyverse, a complete and robust package and essential tool for data science with R.

Summary

Transformations are the core of data wrangling. Datasets are almost like living organisms that change and evolve during the wrangling process, being shaped by the transformations, which, by the way, are driven by the analysis requirements.

In this chapter, we learned about the main transformations for data wrangling in R. We started with slicing and filtering, two great functions for zooming in to a piece of the dataset for deeper analysis. Then we moved on to grouping and summarizing, the dynamic duo of the transformations, where one gathers the data into groups and the other summarizes the essence of the group in a single number or statistic. Replacing and filling was the next section, where we learned about solutions to replace values such as ? with NA, followed by functions to fill NA values with the mean for numeric variables and with the most frequent value for categorical variables.

The section about arranging data covered the use of the order() function to order data in ascending or descending order, facilitating the understanding of the data.

Next, we learned that datasets contain more information than they appear to. If we combine, split, extract, or calculate with data, it is possible to create new variables for better exploration. Yet another way to stretch the dataset is by binding new observations or new variables, topics covered in the section about binding data.

Finally, we covered the essentials of data.table and the enhanced version of data.frames in R, which provides the feature of grouping data. During the chapter, there were a couple of use cases that would not be possible without this library.

I will see you in the next chapter, where we will go over these transformations in a more modern and dynamic way using **tidyverse**.

Exercises

1. What is the difference between slicing and filtering?

2. Describe grouping and summarizing.

3. What function is used to replace all the patterns in a variable?

4. What function drops the missing values from the entire dataset, and when should we use it?

5. What is the percentage of NA values that is OK to drop from a dataset?

6. Describe the main benefit of arranging data.

7. Write a group by command with data.table.

Further reading

- Dua, D. and Graff, C. (2019). UCI Machine Learning Repository [https://archive.ics.uci.edu/ml/datasets/Adult]. Irvine, CA: University of California, School of Information and Computer Science.

- Slicing in R: https://tinyurl.com/yppamu4k

- Arranging data in R: https://tinyurl.com/4e97mvjh

- Introduction to data.table: https://tinyurl.com/4v2kta3e

- The difference between the sub() and gsub() functions: https://tinyurl.com/4a5pbrye

- The nafill function in R: https://tinyurl.com/mk9pzeju

- Code for this chapter on GitHub: https://tinyurl.com/ywpjj329

8
Transformations with Tidyverse Libraries

The journey through data wrangling is still at its core. We have just finished studying the major transformations from the perspective of the built-in functions of base R and counting on the support of **data.table** library.

We saw how easy it was to reach the solution for some of those transformations, without needing to load extra libraries. However, as the problems get more complicated, the basic functions will not be able to provide a sufficiently clean and fast solution. The code will get busier and will probably underperform as the size of the dataset increases. For complex cases, there are several libraries built for R language that can help us to get through most problems with better performance and clean code. If you are interested in comparison times between base R, data.table, and tidyverse, refer to this page (`https://tinyurl.com/2udfcvx2`), where the author compares the most common tasks using the three libraries.

Speaking of data wrangling, I believe that the **tidyverse** package is the most robust tool to deal with data using R. Version 1.0.0 appeared around 2016 (`https://www.rstudio.com/blog/tidyverse-1-0-0/`) as a set of libraries designed to work in harmony and became a one-stop-shop for data science with R. In addition to its clean syntax, it was easier to write and read. Nowadays, some of the libraries comprised in this set constantly figure in the top five most downloaded software from CRAN, the R project repository (`https://tinyurl.com/25wv7dwu`).

Tidyverse will load eight core libraries at once, namely **ggplot2**, **tibble**, **tidyr**, **readr**, **purrr**, **dplyr**, **stringr**, and **forcats**. Other libraries act as assistant parts of the package, dealing with more specific types of objects, thus they are not automatically loaded with a single command. For example, **lubridate**, used in *Chapter 6*, to deal with dates and times, needs to be separately loaded in.

For those more familiar with **SQL**, you will notice many similarities between this language and some functions from the package. It is even possible to read many commands from `dplyr`, for example, as if they have just come out of SQL code.

In this chapter, we will cover the following main topics:

- What is tidy data?
- Slicing and filtering
- Creating new variables
- Joining datasets
- Reshaping a table
- Do more with tidyverse

Technical requirements

Dataset: We will use the Census Income dataset (`https://archive.ics.uci.edu/ml/datasets/Adult`) for this chapter.

All the code can be found in the book's GitHub repository: `https://github.com/PacktPublishing/Data-Wrangling-with-R/tree/main/Part2/Chapter8`. The package to be used can be installed and loaded using the following single commands:

```
install.packages('tidyverse')
library(tidyverse)
```

Figure 8.1 shows the message displayed once the package is loaded. Be aware that there are some red-colored warning messages displayed once you load packages in R, but they are not errors. Those are just to alert you to what version of the R language is used to build their current versions.

```
Console   Terminal ×   Jobs ×
R  R 4.1.0 · ~/
> library(tidyverse)
-- Attaching packages ----------------------------------
âš ggplot2 3.3.5      âš purrr   0.3.4
âš tibble  3.1.6      âš dplyr   1.0.8
âš tidyr   1.2.0      âš stringr 1.4.0
âš readr   2.1.2      âš forcats 0.5.1
-- Conflicts -------------------------------------------
x dplyr::filter() masks stats::filter()
x dplyr::lag()    masks stats::lag()
Warning messages:
1: package 'tidyverse' was built under R version 4.1.2
2: package 'ggplot2' was built under R version 4.1.2
3: package 'tibble' was built under R version 4.1.2
4: package 'tidyr' was built under R version 4.1.2
5: package 'readr' was built under R version 4.1.2
6: package 'purrr' was built under R version 4.1.2
7: package 'dplyr' was built under R version 4.1.2
8: package 'stringr' was built under R version 4.1.2
9: package 'forcats' was built under R version 4.1.2
```

Figure 8.1 – The tidyverse package will load eight libraries. The warning messages are not errors

In the next section, let's remind ourselves about tidy data and why it is important for data wrangling with tidyverse.

What is tidy data

To tidy something means to arrange it, to put it in order. Consequently, tidy data means that our data has a specific order and should follow a set of rules to be considered ready to be worked.

A dataset can be arranged in different ways. For those that, like me, worked for many years with Microsoft Excel, at first sight, a tidy dataset may seem odd, as there will be plenty of repeated cells. Many datasets I worked with in MS Excel had the same measurement split among many columns. A classic example of that is the monthly reports that bring the first columns as the descriptive part of the data (for example, product, profit, and loss), and the values refering to them are shown in one column each month.

Product	Jan	Feb	Mar
Toy	$ 2310	$ 1240	$ 1809
Computer	$ 2788	$ 2342	$ 2002

Figure 8.2 – Example of dataset not in Tidy format

The table from *Figure 8.2* is comfortable to look at but not useful for an algorithm or a programming language. If you try to determine what is the best month for sales, it will require more coding because the computer will not know what a month is. There is no such variable.

Let's step back for a minute to develop a little more intuition about datasets. A dataset is a structured way to arrange data. Every dataset has a subject: cars, stores, medical patients, numerical IDs, and a group of many observations of that subject will compose a dataset. Knowing that, R looks at a dataset and understands each row as one observation at a specific time when it made a measurement about that observation. In the same way, each column is read by R as a different measurement of an observation.

So, we can conclude that R wants to read an organized dataset that has no openings for ambiguous interpretation. There will be one observation per row, one measurement (or variable) per column and one value per cell. This is tidy data.

Back to the example table from *Figure 8.2*, the product is the subject, and it was measured for the sales amount each month. Do you agree that we have one variable, that is, sales, for three different points in time? For that reason, the data is not tidy. To make that data tidy, we must gather all the sales numbers under the same column, associating them with the month and product they pertain to, as seen in the table in *Figure 8.3*.

Product	Month	Sales
Computer	Jan	$ 2788
Computer	Feb	$ 2342
Computer	Mar	$ 2002
Toy	Jan	$ 2310
Toy	Feb	$ 1240
Toy	Mar	$ 1809

Figure 8.3 – Example of a tidy dataset

The dataset looks better from a programming logic standpoint. There is one subject per row: the **product** variable; one point in time per row: the **month** variable; and one sales number by row: the **sales** variable.

The tidyverse libraries carry many transformation functions, including some to transform the dataset to the tidy format, as we shall see in this chapter. To get started, let's go over one more important concept to take advantage of in this package: the forward pipe operator.

The pipe operator

The forward pipe operator, `%>%`, or just pipe for short, is a function from **magrittr** loaded automatically with tidyverse. It enables developers to create chain operations, like a pipe where you input the object in one end and describe the steps to be taken while transforming that input, resulting in a transformed object at the other end.

An analogy that I like to make when I think about pipes is to think about a conversation. Imagine you are chatting with the computer:

- *Developer (D)*: Hey, computer, look at this data frame object. Let's call it `my_data`.
- *Computer (C)*: OK. Understood.
- *D:* Take `my_data` and filter where column 2 is higher than 10.
- *C:* Sure, it's done.
- *D:* Now, can you please group it by column 1 values?
- *C:* Group what?
- *D:* `my_data`.
- *C:* Oh! Of course.
- *D:* And finally, can you calculate the mean and sort it in decreasing order?
- *C:* Sort what?

- *D:* `my_data...`

- *C:* Oh, I sure can.

Notice that the developer had to keep repeating to the computer the name of the object to which the transformations should be applied. This is the same as programming without the pipe operator. But, on the other hand, if you use it, the data will flow in a pipeline where the resulting transformed data from one function becomes the input to the next until the pipeline is completed.

Imagine a table with two columns (`col1` and `col2`) and then examine the following code snippets to see the comparison between programming with and without pipes.

Without pipes	With pipes

```# Data frame```
 ```my data```
 	
 ```# Filter```
 ```my data <- filter```
 ```(my_data, col2 > 10)```
 	
 ```# Group```
 ```my data <-```
 ```group_by(my_data, col1)```
 	
 ```# Summarize```
 ```my data <-```
 ```summarise(my_data, avg=```
 ```mean(col2)```
 	
 ```# Sort```
 ```my data <-```
 ```arrange(my_data,```
 ```desc(avg))```
	```my data %>%```
	 ``` filter(col2 > 10)```
	 ``` group by(col1) %>%```
	 ``` summarise(avg=mean(col2))```
	 ```%>%```
	 ``` arrange(desc(avg))```

Figure 8.4 – Comparison of code snippets without versus with the pipe operator

The comparative table from *Figure 8.4* clearly shows the advantages of working with pipes, in terms of the number of lines of code saved and in the number of times that you have to overwrite your object. Piping also makes the code easier to read and understand, therefore, we will adopt the use of it from now on in this book.

It is valid to say that we won't repeat the whole theory about transformations, but we'll focus more on examples. The dataset to be used is the same one used in *Chapter 7*, the *Adult Data Set (Census Income)* from UCI Repository (`https://archive.ics.uci.edu/ml/datasets/Adult`).

We can load it directly from the repository on the web, using the main function from the `readr` library, as shown in the following code:

```
# Define column names
header <- c("age", "workclass", "fnlwgt","education",
"education_num", "marital_status", "occupation",
"relationship", "race", "sex", "capital_gain", "capital_
loss","hours_per_week", "native_country", "target")
# Load the dataset to RStudio
df <- read_csv("https://archive.ics.uci.edu/ml/machine-
learning-databases/adult/adult.data", col_names = header ,
trim_ws = TRUE)
```

As the dataset does not come with a header row, the `header` vector was created and provided to the `col_names` parameter, along with the `trim_ws=TRUE` parameter, as we know that there are some unwanted leading white spaces in the categorical variables that need to be trimmed. And it is also good to register that tidyverse will always return a tibble object instead of a data frame.

In the following sections, let's start revisiting the list of major transformations in data wrangling, but now performed with tidyverse.

Slicing and filtering

Slicing and filtering a dataset are two similar ways to zoom in on a desired part of the data. In tidyverse, the `dplyr` library deals with the most common data wrangling tasks. Slicing and filtering are among those tasks, as well as the `select()` function, as we will see.

Slicing

As discussed in *Chapter 7*, slicing cuts out unwanted parts of the dataset, returning just part of the rows and/or columns. There is more than one way to slice a dataset, and we will learn the more interesting functions to do that, starting with the most basic one, as follows. In the code, when we use `.`, it means we are considering everything from the object that precedes it, which is `df`, followed by a slicing notation. Rows 1 to 5 and columns 1 to 4:

```
# Slicing rows 1 to 5, columns 1 to 4.
df %>% .[1:5, c(1:4)]
```

The result is shown in *Figure 8.5*.

A tibble: **5 x 4**

age	workclass	fnlwgt	education
<dbl>	<chr>	<dbl>	<chr>
39	State-gov	77516	Bachelors
50	Self-emp-not-inc	83311	Bachelors
38	Private	215646	HS-grad
53	Private	234721	11th
28	Private	338409	Bachelors

Figure 8.5 – Sliced dataset

To provide summary notes about the previous code: the dataset object was mentioned only once; we used pipe to link both commands; to slice this package, we needed to use a point before the slicing notation, and the returned object is a tibble.

We can use `slice_min()` or `slice_max()` to fetch the top or bottom observations based on a variable and the percentage we need to see. For example, to see the observations with the youngest 10% of people or if we want to return the oldest 30%, this is the code:

```
# Slicing with slice_min() and slice_max()
df %>% slice_min(age, prop=0.10)
df %>% slice_max(age, prop=0.30)
```

Data science is very closely related to statistics, so data scientists often work with samples. With `dplyr`, extracting a sample of the dataset is fairly simple using the `slice_sample()` function. The next code snippet is used to randomly collect a sample of 10 observations with replacement, meaning that the same observation could be picked more than once:

```
# Slice sample
df %>% slice_sample(n=10, replace=TRUE)
```

Let's move on to study how to filter datasets.

Filtering

Filtering is more commonly applied to one of the variables, using a condition to filter out the data you don't want to see and returning all the variables for the filtered observations.

In the sequence, let's look at the `filter()` function from the `dplyr` library:

```
# Filtering age over 30 years old
df %>% filter(age > 30)
```

For the results, see *Figure 8.6*.

A tibble: 21,989 x 15

age	workclass	fnlwgt	education	education_num
<dbl>	<chr>	<dbl>	<chr>	<dbl>
39	State-gov	77516	Bachelors	13
50	Self-emp-not-inc	83311	Bachelors	13
38	Private	215646	HS-grad	9
53	Private	234721	11th	7
37	Private	284582	Masters	14

Figure 8.6 – A filtered dataset on observations with age over 30 years old

Once again, we have a tibble object returned, but this time containing all 15 variables in it.

Using the `select()` function allows the selection of a set of variables, making filtering more like slicing. If we only want to filter the data on people over 30 years old to look at their marital status, we use the code in the following sequence:

```
# Filter age > 30 and selecting age and marital_status
df %>%
    filter(age >30) %>%
    select(marital_status, age)
```

The code displays the results shown in *Figure 8.7*.

A tibble: 21,989 x 2

marital_status	age
<chr>	<dbl>
Never-married	39
Married-civ-spouse	50
Divorced	38
Married-civ-spouse	53

Figure 8.7 – dplyr functions filter and select used together

Using the `select()` function empowers the developer to choose the order to display the variables, and select which ones to show. Also note that it is good practice to break the code into different lines after the `%>%` pipe signal, making it more organized and easier to read.

Still on the subject of filtering, there is the distinct() function, frequently used in the SQL world to remove duplicate entries. When analysts are still learning about the data, they can use this resource to know how many groups there are within a variable. For example, with the Census data, if we don't know how many genders are being considered, we can look for distinct values, which can be written as follows:

```
# Distinct - removing duplicates
df %>% distinct(sex)
```

This code returns the unique entries for the selected variable (sex), which are Male and Female. But it can return unique observations for the entire dataset if you leave the parenthesis empty.

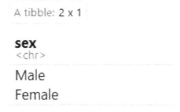

Figure 8.8 – Unique entries for the sex variable

Moving on, the following section covers grouping and summarizing data.

Grouping and summarizing data

Grouping and summarizing are two complementary functions. Generally, they will be used together, as there is not much use in grouping a dataset and not calculating anything or using the groups for a purpose. That is when summarizing plays the important role of transforming the data from each group into a summary or a number that we can understand.

In the business world, requests such as the average number of sales by store, the median number of customers by day, the standard deviation of a distribution, and many other examples, are part of the routine of a data scientist. These tasks can be performed using the group_by() and summarise() functions from dplyr.

Starting with the group_by() function, observe that it alone cannot bring much value:

```
# group by not summarized
df %>% group_by(workclass)
```

Here is the result.

A tibble: 32,561 x 15   Groups: workclass [9]

age <dbl>	workclass <chr>	fnlwgt <dbl>	education <chr>
39	State-gov	77516	Bachelors
50	Self-emp-not-inc	83311	Bachelors
38	Private	215646	HS-grad
53	Private	234721	11th
28	Private	338409	Bachelors
37	Private	284582	Masters

Figure 8.9 – Dataset grouped but not summarized

We can see in *Figure 8.9* that it worked because there is a box displaying the number of groups created (9) on the upper center of the picture, which is a great feature from tibble, by the way. But if we want to calculate the average age of each group of the workclass variable, the group_by() function alone will not help us.

On the flip side, the summarise() function can bring us some useful information even without being attached to groups. Look at the following code. Let's calculate the average age of the people in this dataset:

```
# summarise without group_by returning average age of the
dataset
df %>% summarise( age_avg = mean(age) )


age_avg
<dbl>
38.58165
```

Summarizing is to bring the data down to a single number, so we received the average age of 38.58 years old for the *Census Income dataset*. It is useful information, but we could probably use much simpler code from base R to calculate that, such as mean(). So we can see that using group_by() or summarise() separately is not as powerful as using them combined. When grouping and summarizing are glued together, they work at their best, allowing analysts to answer business questions and extract good insights from the data.

We want to know what workclass has the oldest people working in it, on average. That insight can lead us to understand how many years the people in each area are still expected to work and contribute until they start to retire. Let's calculate that here:

```
# Group by workclass and summarize mean age
df %>%
```

```
group_by(workclass) %>%
summarise(age_avg = mean(age))
```

As a result, the software displays the `workclass` groups and the summarized age average by group.

People without pay are the oldest, on average, followed by the self-employed, which could be concerning if they are not saving enough money on their own. They are closer to retirement than the other groups.

A tibble: 9 x 2

workclass <chr>	age_avg <dbl>
?	40.96024
Federal-gov	42.59063
Local-gov	41.75108
Never-worked	20.57143
Private	36.79759
Self-emp-inc	46.01703
Self-emp-not-inc	44.96970
State-gov	39.43606
Without-pay	47.78571

Figure 8.10 – Group by workclass and summarize by age

Figure 8.10 has a simple grouping structure, with only one variable being grouped (`workclass`) and another one being summarized (`age`). That will not always be the case, naturally. When grouping by more than one variable, `dplyr` returns an object called `grouped_df`, which at first sight may not bring trouble. However, a few functions and libraries don't get along too well with that type of object. Ergo, I recommend you get into the habit of using the `ungroup()` function, which resolves this problem.

We can compare two pieces of code and look at their resulting object types. The first one will result in `grouped_df`:

```
# Returns object grouped_df
workclass_groups <- df %>%
  group_by(workclass, sex) %>%
  summarise(age_avg = mean(age) )
```

The second, using `ungroup()`, returns a `tbl_df`:

```
# Returns object tibble
workclass_ungrouped <- df %>%
```

```
group_by(workclass, sex) %>%
summarise(age_avg = mean(age) ) %>%
ungroup()
```

From the preceding lines of code, the objects displayed on the R **Environment** tab will be as seen in *Figure 8.11*.

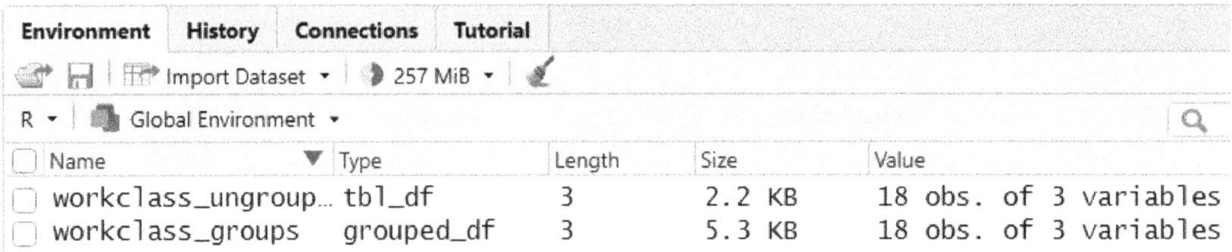

Name	Type	Length	Size	Value
workclass_ungroup...	tbl_df	3	2.2 KB	18 obs. of 3 variables
workclass_groups	grouped_df	3	5.3 KB	18 obs. of 3 variables

Figure 8.11 – Objects returned with and without the use of the ungroup() function

It is also valid to point out that the ungrouped tibble requires less memory space than the other object.

We have covered the majority of what is important about grouping and summarizing, but I would like to spend a few more lines on the summary functions available. It is not all about `mean`, `median`, and `sum`. There are other useful functions to be used within `summarise()` to be listed subsequently. The explanations are in the comments preceding each code snippet:

```
# n() shows the count of observations in each group
df %>% group_by(workclass) %>% summarise(n())


# n_distinct() shows the count of unique observations in each
group
df %>% group_by(workclass) %>% summarise(n_distinct())


# sum(!is.na()) shows the count of Non NA observations
df %>% summarise(sum(!is.na(workclass)))


# first() shows the first age value in each group
# Similarly, you can use last() or nth()
df %>% group_by(workclass) %>% summarise(first(age))


# quantile() shows the top number of the quantile percentage
chosen
# Here, 50% of the age observations are under what value by
group
```

```
df %>% group_by(workclass) %>% summarise(quantile(age, 0.5))

# sd() shows standard deviation of a variable
df %>% group_by(workclass) %>% summarise(sd(capital_gain))

# Across function: applies function to the selected columns
(ex. mean)
df %>% select(1,3,5,11,12,13) %>%
   summarise(across(everything(), mean))
```

The next section will cover how to replace values and fill in missing or faulty data to advance further into data wrangling transformations. For that, we will require the tidyr library.

Replacing and filling data

A dataset can and certainly will be acquired with imperfections. An example of imperfection is the use of the ? sign instead of the default NA for missing values for the *Census Income dataset*. This problem will require the question mark to be replaced with NA first, and then filled with another value, such as the mean, the most frequent observation, or using more complex methods, even machine learning.

This case clearly illustrates the necessity of replacing and filling data points from a dataset. Using tidyr, there are specific functions to replace and fill in missing data.

First, the ? sign needs to be replaced with NA, before we can think of filling the missing values. As seen in *Chapter 7*, there are only missing values for the workclass (1836), occupation (1843), and native_country (583) columns. To confirm that, a loop through the variables searching for ? would be the fastest resource:

```
# Loop through variables looking for cells == "?"
for (variable in colnames(df)){
  print(
    paste( variable,
          dim( df[df[variable]=="?", variable])[1] )
  )
}
```

The last code snippet has a for loop structure looking at each variable for cells equal to ?, adding those observations to a subset. As we know that the result of slicing a tibble is another tibble object with rows and columns, then calling the dim() function returns a vector of the dimensions of that subset. Slicing its first position will return the number of rows with ? that was pasted with the variable name and printed on the screen. The returned values would be as follows for the variables with NA present and 0 for the others:

```
[1] "workclass 1836"
[1] "occupation 1843"
[1] "native_country 583"
```

Once we know which columns must have the values replaced, we can move forward using the next code snippet. It takes the data frame and overwrites the `workclass`, `occupation`, and `native_country` variables using the `mutate` function in combination with `replace()`, which replaces the interrogation marks with NA:

```
# Replacing values "?" with NA, saving in a new dataset
variable
df_replaced <- df %>%
  mutate(workclass = replace(workclass, workclass == "?", NA),
        occupation = replace(occupation, occupation == "?",
NA),
        native_country = replace(native_country, native_
country == "?", NA))
```

The output is the `df_replaced` tibble with NA values instead of ?. Please see *Figure 8.12* for the output.

▲	age	workclass	fnlwgt	education	education_num	marital_status	occupation
26	56	Local-gov	216851	Bachelors	13	Married-civ-spouse	Tech-support
27	19	Private	168294	HS-grad	9	Never-married	Craft-repair
28	54	*NA*	180211	Some-college	10	Married-civ-spouse	*NA*

Figure 8.12 – NA values observed on row number 28

The code combining `mutate` and `replace` works perfectly fine, but it requires four lines. There is a single function created especially for that kind of replacement called `na_if()`, from `dplyr`. It works using the logic that it replaces the pattern with NA if the value matches it, which in our case would be the pattern ?. Let's see it in action:

```
# Replacing values "?" with NA
df_replaced <- df %>% na_if("?")
```

It works like a charm, and the result is exactly the same as seen in *Figure 8.12*.

Handling the missing values is easier than replacing values. That is because tidyr has neat functions for that task. To keep the focus on data wrangling, we will go over some simple methods to handle

NAs, such as dropping them, filling them with the previous or next valid value, or replacing them with a statistic value, such as the mean value.

The first approach covers filling in missing values with the previous or next valid cell. Look at the solution in the following sequence:

```
# Fill NA values with last or next valid value
df_replaced %>% fill(workclass, occupation, native_country,
                     .direction= "down" )
```

Using the `fill()` function, it is possible to choose the direction of the fill. `.direction= "down"` uses the last valid cell to fill the values down.

	age	workclass	fnlwgt	education	education_num	marital_status	occupation
26	56	Local-gov	216851	Bachelors	13	Married-civ-spouse	Tech-support
27	19	Private	168294	HS-grad	9	Never-married	Craft-repair
28	54	Private	180211	Some-college	10	Married-civ-spouse	Craft-repair

Figure 8.13 – Values on row 28 were filled with the last valid value, from row 27

Figure 8.13 shows that the NA values from row 28 were filled down, repeating the value from row 27. If `.direction= 'up'` is used instead, the fill value would come from row 29, as long as it is not another NA. It looks at the last or next valid value to fill the gaps.

When the variable is categorical, a common approach to filling in missing values is to use the most frequent value, the one that appears more frequently in our data. Using that, we will have more chances to add it correctly, purely by probability. Every change in a dataset, though, is a decision that the data scientist has to make and it will certainly affect the result. As already mentioned, there are newer ways to do that filling using the power of machine learning, but for educational purposes and to keep to the scope of this book, let's see the code to fill the values with the most frequent observation by variable.

First we created two intermediate variables with the most frequent values, using the same code for this task as we did in *Chapter 7*:

```
# Finding the most frequent entry
m_freq_workcls <- names(table(df$workclass)[which.
max(table(df$workclass))])
m_freq_occup <- names(table(df$occupation)[which.
max(table(df$occupation))])
```

Then, we take the `df_replaced`, which is the data frame with ? replaced with NA and added the most frequent values calculated in the previous step. Those are added to the `workclass` and `occupation` columns using the `replace_na()` function. This function works with a list of variables where it is required to pass the variable name and the replacement value. Remember to provide the function with the same type of data from each variable. Also, notice that we saved the result in a new variable name, `df_no_na`:

```
# Replace NA with the most frequent value
# and save the dataset in a new variable name
df_no_na <- df_replaced %>%
  replace_na(list(workclass= m_freq_workcls,
                  occupation= m_freq_occup) )
```

I recommend that you create a new variable name every time a major transformation is done, such as changing or deleting values. That keeps you backed up in case there is a need to go back one step.

Finally, there is the `drop_na()` function, which is the easiest way to deal with missing values by simply removing them. This does not mean it is the right or wrong answer, just the easiest solution. Although, good practice tells us to observe whether the total number of NAs is lower than 5% of the total observations. For `workclass` and `occupation`, the percentages were a little over 5%, so we decided to fill in the missing data with the most frequent value. However, for `native_country`, it is less than 2%, so let's go ahead and drop the NAs left in the dataset now:

```
# Drop NAs
df_no_na <- df_no_na %>% drop_na()
```

Nice and easy, they are dropped, and the dataset is cleaned of missing values. With that, it is time to continue our studies. Coming up, we will go over arranging data functions.

Arranging data

Arranging data is useful to create a rank, making the dataset ordinated. The orders can be from low to high values, also known as increasing order, as well as from high to low or decreasing order. In RStudio, visualizing a dataset using the software's viewer pane already allows the analyst to arrange the data with the click of a button. Just like many dynamic tables, if you click on a column name, that variable becomes ordered. For simply eyeballing it, the feature is terrific, but for programming purposes, it won't have any effect. You will have to take advantage of the `arrange()` function from `dplyr`.

The most basic ways to arrange a dataset are by running the succeeding pieces of. First, let's try arranging by increasing order:

```
# Arrange data in increasing order
df_no_na %>% arrange(native_country)
```

Next, arranging in decreasing order:

```
# Arrange data in decreasing order
df_no_na %>% arrange( desc(native_country) )
```

Notice that adding the `desc()` function changes the order direction.

However, arranging data is more useful when the data is grouped and summarized. In the last chapter, we saw a comparison between a grouped and ordered dataset and an unordered one. Clearly, it is quicker to absorb the information from the ordered data.

Now that our dataset is clean of missing values, let's exploring it further. We want to know whether there is a gain in the net gain for those people who achieved higher education levels. Let's learn how to do that using tidyverse functions. We can group the data by the education level and calculate the average of the net gain (`capital_gain - capital_loss`). In addition, we can also calculate how many observations each group has to see how significant that is to the dataset. Finally, we will arrange the result in decreasing order:

```
# Group and order average net gain by education level
df_no_na %>%
   group_by(education) %>%
   summarise(count=n(),
           avg_net_gain= mean(capital_gain - capital_loss))
%>%
   arrange( desc(avg_net_gain) )
```

The previous code snippet returns the data shown in *Figure 8.14*.

education <chr>	count <int>	avg_net_gain <dbl>
Prof-school	559	10259.63327
Doctorate	390	4672.28205
Masters	1674	2357.94146
Bachelors	5210	1608.20077
Preschool	50	848.54000
Assoc-voc	1366	652.77965
Some-college	7187	534.44734
HS-grad	10368	500.05498
Assoc-acdm	1055	459.03223
10th	921	354.40608

Figure 8.14 – Data grouped and arranged in decreasing order

Looking at the ordered table in *Figure 8.14*, we can quickly understand it and elaborate on our takeaways. It presents some association between the education level with the average net gain amount of a person. At the top, we can observe people with completed studies from a professional school or with a doctorate, a master's, or bachelor's degree. At the bottom, the lowest gains are from people with lower education levels. An interesting result is people with only preschool completed appearing in the top five. At a second look, though, there are only fifty observations like that. However, that could be something to explore further in a real project.

The following section is about creating new variables to be able to develop a better analysis.

Creating new variables

Creating new variables can be useful for data scientists when they need to analyze something that is not present in the data as it was acquired. Common tasks to create new data are splitting a column, creating a calculation, encoding text, and applying a custom function over a variable.

We went over some good examples of column splitting in this book, such as a datetime split. Now, to illustrate the `separate()` function from tidyr, the example to be used is based on the *Census Income dataset*. Look at the `target` column: it has values such as <=50k and > 50k. Let's say we wanted to separate only the > or <= signs and put them in a separate column; here is how to do that:

```
# Split variable target into sign and amount
df_no_na %>% separate(target, into=c("sign", "amt"), sep="\\b")
```

We took the dataset clean of NAs and separated the `target` column into two new variables: `sign` and `amt`. To accomplish that, the `sep` parameter takes a *regexp* pattern as input, for which we used a boundary pattern, resulting in *Figure 8.15*.

capital_loss <dbl>	hours_per_week <dbl>	native_country <chr>	sign <chr>	amt <chr>
0	52	United-States	>	50K
0	50	Germany	<=	50K
0	40	United-States	<=	50K
0	40	United-States	<=	50K
0	40	United-States	>	50K
0	44	United-States	>	50K

Figure 8.15 – The target column is separated into two new columns

Observe that the `target` column is gone since the `separate()` function has the `remove=TRUE` parameter by default. So, be aware of that and use `remove=FALSE` if you want to keep the original column.

If we can split data, we can unite it too. The `unite()` function can be helpful if we need to create a unique identifier. To illustrate the concept, let's suppose you work at a retail business where you have products being sold in stores. We all know that, usually, the same product is sold in more than one store. Therefore, to create a unique identifier of product 1 sold at store 10, we can gather both variables, creating the value 1-10, for example.

Transferring the concept to our dataset, let's create a `description` column by uniting `sex`, `race`, and `age`. Just as a reminder, in the subsequent code, we are not actually changing the dataset since the result is not being assigned with `<-` to a variable name:

```
# Unite variables sex, race and age
df_no_na %>%
  unite(sex, race, age, col="description", sep="_",
remove=FALSE)
```

This time, as the `remove=FALSE` parameter, the original variables are kept. The separator is set by the `sep` parameter, where we used the underscore mark. See the result in *Figure 8.16*.

description <chr>	age <dbl>	workclass <chr>	fnlwgt <dbl>	education <chr>
Male_White_39	39	State-gov	77516	Bachelors
Male_White_50	50	Self-emp-not-inc	83311	Bachelors
Male_White_38	38	Private	215646	HS-grad
Male_Black_53	53	Private	234721	11th
Female_Black_28	28	Private	338409	Bachelors

Figure 8.16 – Column description created as a union of sex, race, and age

It worked as expected.

The mutate function

The `mutate()` function is a versatile and widely used method from `dplyr` and was designed to create new variables. As you might have noted, the tidyverse functions use intuitive verbs that tell us or at least give us an idea of the transformation they perform. Thus, when there is a need to create a mutation to the dataset, or an addition, I should say, this function comes in very handy.

The syntax follows the standard of the others from tidyverse, requiring a dataset and the variable to be added, which can be a vector, a statistic, a custom calculation, or other custom options that your creativity can think of. See a generic example of its use in the following snippet:

```
mutate( df, new_var= c(1,2,3,4,5) )
```

In this case, provided that our dataset has five rows, a column named `new_var` would be added with the numbers from the vector as its values. Certainly, `mutate` can be used with the pipe notation, eliminating the need to pass the dataset as the object to `mutate()`.

Revisiting the same example from the last chapter, where we created a new variable that calculated the net gain (`capital_gain - capital_loss`) and calculated a tax fee of 10% if the amount was over 15,000, the following code shows how to perform that activity using `mutate()`:

```
# Tax variable creation
df_no_na %>%
  mutate(total_gain = capital_gain - capital_loss,
         tax= ifelse(total_gain >= 15000,
                     total_gain *0.1,
                     0)  ) %>%
  arrange(desc(tax))
```

Explaining what was done, we took the data without NAs and used `mutate()` to create two new variables. The first one is `total_gain`, a calculation of the net gain for each observation, and the second variable is named `tax`, created out of a condition that if the `total_gain` is over or equal to 15,000, a tax fee of 10% is applied, otherwise, not.

This extraction of the result shows both new columns created.

native_country <chr>	target <chr>	total_gain <dbl>	tax <dbl>
United-States	>50K	27828	2782.8
United-States	>50K	27828	2782.8
United-States	>50K	27828	2782.8
United-States	>50K	27828	2782.8
United-States	>50K	27828	2782.8

Figure 8.17 – New columns created out of a custom calculation

In the previous example, we saw `mutate()` being used with `ifelse()`, showing how we can customize new variables using this powerful function. Another nice combination to be studied is using `mutate()` and `recode()`, working as a mapping to transform values. Going back to the *Census Income dataset* once again, the variable target can change the values to something more comfortable to read. There are two unique values. Let's change `<=50K` to `under` and `>50K` to `over`:

```
# Change values from target to over or under
df_no_na %>%
  mutate( over_under=
            recode(target, '<=50K'='under', '>50K'='over')    )
%>%
  select(target, over_under)
```

The code takes the dataset and creates a mutation by creating a new column named `over_under`, which will recode the values from symbols on the left-hand side of the equal signs to words on the right-hand side. Finally, we select only the original and the newly created columns. It will display the table in *Figure 8.18*.

target <chr>	over_under <chr>
<=50K	under
<=50K	under
<=50K	under
<=50K	under
<=50K	under
<=50K	under
<=50K	under
>50K	over
>50K	over
>50K	over

Figure 8.18 – The mutate and recode functions working together to map values

To close this section, let's go over one more example. Observe the next code snippet. We get our dataset and call the `mutate()` function to create two new variables: `age_avg` and `over_under_age_avg`. The first one is just the average age number repeated for the whole column, for comparison purposes. The second new variable was created with the `cut()` function over age. We input a vector of interval cuts, beginning at zero, with a cut at the mean value and the final cut at the maximum value. So, there are two intervals: from 0 to the mean and from mean to the maximum. The second vector gives names to each interval; thus, everything that falls within interval one is marked as `Lower than avg`, and what falls in interval two is marked as `Above the avg`. To finish, we selected the columns to display:

```
# Observations over or under the average
df_no_na %>%
  mutate( age_avg = mean(age),
          over_under_age_avg= cut(age,
                        c(0, mean(age), max(age)),
                        c('Lower than avg', 'Above the
avg')) ) %>%
  select(age, age_avg, over_under_age_avg)
```

After running the code, R displays the table you see in *Figure 8.19*.

age <dbl>	age_avg <dbl>	over_under_age_avg <fctr>
39	38.57902	Above the avg
50	38.57902	Above the avg
38	38.57902	Lower than avg
53	38.57902	Above the avg
28	38.57902	Lower than avg
37	38.57902	Lower than avg
49	38.57902	Above the avg
52	38.57902	Above the avg
31	38.57902	Lower than avg
42	38.57902	Above the avg

Figure 8.19 – The mutate and cut functions were used to create a classification of the observations

Figure 8.19 demonstrates that creativity drives what you can do while creating new variables. Now we know which person is over or under the average age in this dataset.

I believe that the capabilities of mutate() are limited only by the amount of creativity you have to match functions and create customized new variables. We have seen a couple of compelling use cases that should be enough to create the base for you to develop more complex code using this function. In the *Further reading* section, you will find other use cases.

The continuation of this chapter will examine joining datasets. The concepts to be covered next are closely related to SQL and are an excellent tool for your set of skills.

Joining datasets

Datasets can come from different sources or different tables within the same database or data lake. Many times, those tables are related to each other by **key** columns, which means that you will be able to find a certain column A in table 1 and a column A in table 2 that hold similar information so they can be related to each other using that common key element.

To better explain the join concept, imagine we are engineers from a retail company. Our goal is to store data about transactions from each store, including date, product, descriptions, quantity, and amount. Well, we can put everything in the same table, resulting in a big heavy file that the database will have to deal with every time we want to query some information. Think about that for a moment: it won't be every time that we will need to pull the product description, or store address, for example. Consequently, the optimal solution for that problem is splitting that information into smaller tables related to each other and storing everything in a database.

A popular relational database architecture, also called a schema, is the *star schema* (`https://en.wikipedia.org/wiki/Star_schema`), which divides the data into centralized facts tables and dimension tables. The facts are comprised of metrics and numbers from the observations, while the dimensions are the characteristics and descriptions of the observations.

Back to the retail example, the facts are the sales and quantities, while the dimensions are the stores, product descriptions, color, and size. This way, it is possible to pull only basic information and add whatever extra piece we need from the related dimension tables.

Facts

Date	Store_cd	Product_cd	Quantity	Sales
2022-01-01	1	001	10	$30
2022-01-02	2	002	12	$60
2022-01-03	3	003	9	$45

Dimension Store

Store_cd	Address	City	Open Hours
1	123 This street	South City	7 - 23
2	234 That street	North City	7 - 22
3	456 Main blvd	Central City	6 - 23

Dimension Product

Product_cd	Description	unit_price	unit_measure
001	Soft drink	$3.0	each
002	Frozen snack	$5.0	each
003	Fruit	$5.0	kg

Figure 8.20 – Related tables: facts and dimensions

As seen in *Figure 8.20*, the information was split into three tables. The facts table holds the amounts and the dimension tables have supporting information that probably won't be needed for every query, thus don't need to be handled every time, saving memory and increasing performance. In any case, the information from the dimension tables can be joined to the facts whenever needed using the common key variables highlighted in orange and yellow. That is what we are going to learn next.

Before we start coding, let's learn more about the different types of joins:

Type of Join	Dplyr Function	Illustration
Left join: this returns all the rows from the left table but only the matched rows from the right table. Unmatched values return NA.	`left_join ()`	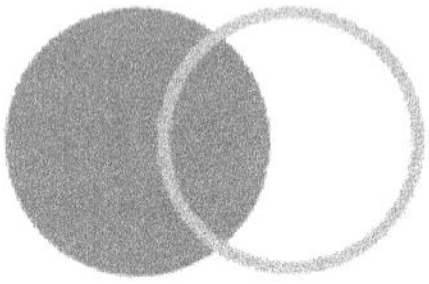
Right join: this returns all the rows from the right table but only the matched rows from the left table. Unmatched values return NA.	`right_join ()`	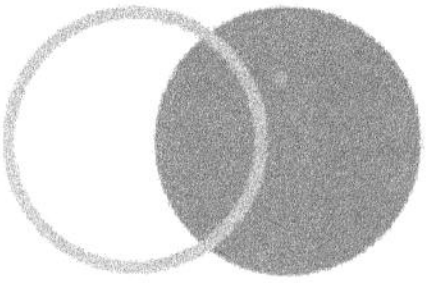
Inner join: this returns only the matched rows from both tables. It is the intersection between the left and right tables. Unmatched values are omitted.	`inner_join ()`	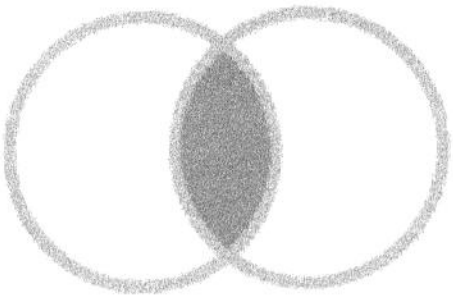
Inner join: this returns all rows from left and right tables, even if not matched.	`full_join ()`	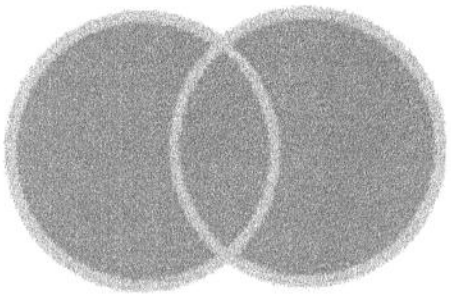
Anti join: this returns rows from the left table that are not in the right table.	`anti_join ()`	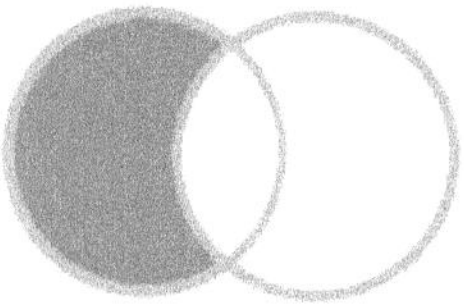

Figure 8.21 – Types of joins

Given that dplyr has a lot in common with SQL, the function names presented in *Figure 8.21* use the same names of the types of joins, making it easier to remember. If you understand the illustrations, then it becomes much easier to learn when to use each function.

Aiming to deepen our knowledge, let's learn by following examples. For the demonstrations in this section, I will create the same tables from *Figure 8.20*, but I will add a couple of rows that don't have a match. You can find the code for this in the files that accompany this book (https://tinyurl.com/yckswc37). The tables are presented in *Figure 8.22*.

Facts				
Date	Store_cd	Product_cd	Quantity	Sales
2022-01-01	1	1	10	$30
2022-01-02	2	2	12	$60
2022-01-03	3	3	9	$45
2022-01-04	4	4	12	$24
2022-01-05	5	5	8	$32

Dimension Store			
Store_cd	Address	City	Open Hours
1	1 main st	Main	7 - 23
2	20 side st	East	7 - 23
3	19 square blvd	West	9 - 21
4	101 first st	North	9 - 21
6	1002 retail ave	South	9 - 21

Dimension Product			
Product_cd	Description	unit_price	unit_measure
1	Soft drink	$3.0	each
2	Frozen snack	$5.0	each
3	Fruit	$5.0	Kg
4	Water	$2.0	each
6	Fruit 2	$4.0	Kg

Figure 8.22 – Tables created for the exercises about the types of joins

Let's begin working on the coding part of the example now.

Left Join

Having the tables from *Figure 8.22* in mind, suppose we want to know what the product's descriptions are. Using left_join(), this can be done easily:

- Left table: sales
- Right table: products

```
# Left join
sales %>%
  left_join(products, by= 'product_cd')
```

The result is as follows.

Description: df [5 x 8]

date <chr>	store_cd <dbl>	product_cd <dbl>	qty <dbl>	sales <dbl>	description <chr>	unit_price <dbl>	unit_measure <chr>
2022-01-01	1	1	10	30	Soft drink	3	each
2022-01-02	2	2	12	60	Frozen snack	5	each
2022-01-03	3	3	9	45	Fruit	5	kg
2022-01-04	4	4	12	24	Water	2	each
2022-01-05	5	5	8	32	NA	NA	NA

Figure 8.23 – The left join joins sales with products

Notice in *Figure 8.23* that the last row has some *NA values*. If we revisit the theory about left joins, remember that it will keep all the observations from the left table (`sales`) and join all the matching observations and variables from the right table (`products`), knowing that when there is no match, the return is a missing value. If you go back to *Figure 8.22*, you will spot that there is no product with code number 5 in the `products` dimension table, therefore, returning an *NA* entry.

This is very good, but we can do even better. Our problem here was to bring only the description of the product, so there is no need to join all the columns from the right table. What we can do is select only the key column `product_cd` because we need something to match `by`, and the `description`, which is what we want to have in our `sales` table. In R code, that becomes the following snippet:

```
# Left join with selected columns from products
sales %>%
  left_join(products[,1:2], by='product_cd')
```

And the result is displayed as follows.

Description: df [5 x 6]

date <chr>	store_cd <dbl>	product_cd <dbl>	qty <dbl>	sales <dbl>	description <chr>
2022-01-01	1	1	10	30	Soft drink
2022-01-02	2	2	12	60	Frozen snack
2022-01-03	3	3	9	45	Fruit
2022-01-04	4	4	12	24	Water
2022-01-05	5	5	8	32	NA

Figure 8.24 – Left join of selected columns from products

Figure 8.24 is cleaner than the previous version, making it easier to read.

Let's move on to right join.

Right join

Right join is simply the opposite of the left join. They work in exactly the same way, but in this case, the observations kept as primary are from the table on the right. We can use the `stores` table on the right and bring the sales amount from the `sales` table. Let's look at the code:

- Left table: `sales`
- Right table: `stores`

```
# Right join
sales %>%
  select(store_cd, sales) %>%
  right_join(stores, by= 'store_cd')
```

Observe that we have selected only the key column to join by and the `sales` variable needed in the result in the upcoming *Figure 8.25*.

store_cd <dbl>	sales <dbl>	address <chr>	city <chr>	open_hours <chr>
1	30	1 main st	Main	7-23
2	60	20 side st	East	7-23
3	45	19 square blvd	West	9-21
4	24	101 first st	North	9-21
6	NA	1002 retail ave	South	9-21

Figure 8.25 – Right join from sales with products

Only the `sales` column is an addition to the table in this case. It is important to mention that the key column `store_cd` was not repeated because the `keep=FALSE` parameter is set by default. Changing it to `TRUE` would cause the equal variables from both tables to repeat. Furthermore, it is possible to perform any type of join using key columns with different names that carry matching information. For example, if we had the column `store_cd` in the left table and `store_code` in the right table, we must use this configuration: `by = c('store_cd' = 'store_code')`.

Inner join

The inner join is commonly used because it filters off the non-matching entries. Let's jump directly to the code:

- Left table: `sales`
- Right table: `stores`

```
# Inner join
sales %>%
  inner_join(stores, by='store_cd')
```

Remember that the `sales` table does not have a store number 6, and the `stores` table does not have a store number 5. Neither will be present in the result as seen in the following figure.

Description: df [4 x 8]

date <chr>	store_cd <dbl>	product_cd <dbl>	qty <dbl>	sales <dbl>	address <chr>
2022-01-01	1	1	10	30	1 main st
2022-01-02	2	2	12	60	20 side st
2022-01-03	3	3	9	45	19 square blvd
2022-01-04	4	4	12	24	101 first st

Figure 8.26 – Inner join of sales with stores

As expected, the table in *Figure 8.26* has only four matching observations.

The opposite version of the inner join is the full join and is what we are about to learn about in the following section.

Full join

The full join will put everything together from both tables, not caring if it is a match or not. The observations that don't match will return NA. If in the previous example, we received only four matching observations, this time, we should expect to see all six values being displayed. In this code snippet, I did not include the by parameter on purpose, so you can see how `dplyr` will handle it:

- Left table: `sales`
- Right table: `stores`

```
# Full join
sales %>% full_join(stores)
```

There will be a message on the screen saying **Joining, by = "store_cd"** and the result will appear as follows.

store_cd <dbl>	product_cd <dbl>	qty <dbl>	sales <dbl>	address <chr>	city <chr>	open_hours <chr>
1	1	10	30	1 main st	Main	7-23
2	2	12	60	20 side st	East	7-23
3	3	9	45	19 square blvd	West	9-21
4	4	12	24	101 first st	North	9-21
5	5	8	32	NA	NA	NA
6	NA	NA	NA	1002 retail ave	South	9-21

Figure 8.27 – Partial screenshot of the full join of sales and stores

Figure 8.27 shows the returned table from the code. As expected, store number 5 does not have information available, and store number 6 does not have facts available, both returning missing values.

Last but not least, the anti-join will be covered in the next subsection.

Anti-join

Anti-join looks at both tables and checks what is in the left table that is not in the right table, and that is what it returns. From the `sales` table, only product number 5 is not in the `products` dimension table:

- Left table: `sales`

- Right table: `products`

```
# Anti-join
sales %>% anti_join(products)
```

Once again, as I didn't include the by parameter, it automatically determined that the `product_cd` key column was the only one present on both sides, and returned the table in *Figure 8.28*.

Description: df [1 x 5]

date <chr>	store_cd <dbl>	product_cd <dbl>	qty <dbl>	sales <dbl>
2022-01-05	5	5	8	32

Figure 8.28 – Anti-join of sales and products. Only product 5 is not present in the products table

As only product 5 was not in the `products` table, the returned data frame has only that one observation.

With this example, we have reached the end of this section about types of joining data. This knowledge is highly valuable, and I recommend that you continue learning from the links provided in the *Further reading* section.

The next section will teach us how to reshape a table, transforming it from a wide to a long format or the other way around.

Reshaping a table

Data frames are composed of rows and columns, and we already know that data that is considered tidy has one observation per row and one variable per column. However, data is everywhere, being generated by different kinds of sensors, machines, devices, and people. Naturally, not all that data will come to you in a perfectly beautiful, tidy format. Many will be the time when you will face messy datasets.

One of the things that can happen is to receive datasets in a wide format. As the name already suggests, this data comes with variables spread along the rows instead of columns, meaning that measurements from the same variable can be displayed on different columns. A common example of a wide formatted dataset is a monthly report with a column for the description of a project, followed by a column with the expense amount for each month, as shown in *Figure 8.29*.

	Expenses		
project	Jan	Feb	Mar
project1	1560	1152	1145
project2	1996	1073	1633
project3	1320	1227	1048

Figure 8.29 – Projects dataset in wide format

A data frame in a wide shape is not suitable for machine learning algorithms, as well as for most of the plotting functions. This happens because the computer expects a single column name by variable to know what we are talking about, not a set of names. A bunch of columns is interpreted as different variables. So, in the case of a plot, we can't tell the computer to use the column names (month names) as the *x* axis, and the dollar amounts as the *y* axis. It expects us to pass the variable for *x*, which would be the names of the months, maybe a variable called `months`, and another variable for *y*, which would be the expenses.

To reshape the data frame from *Figure 8.29* from wide format to long, use the following code:

```
# Wide to Long
df_long <- df_wide %>%
  pivot_longer(cols= 2:7,
               names_to = 'months',
               values_to = 'expenses')
```

We called our dataset `df_wide` and piped it with the `pivot_longer()` function from tidyr, specifying which columns should be reshaped to the long format using the `cols` argument. Column 1, not listed in this argument, will stay fixed as the pivoting point, with the observations being automatically duplicated as needed. Then, what will happen is that the names of each column (*Jan, Feb, Mar, Apr, May, and Jun*) will become one column, and the name of that column is set by the `names_to` parameter. All the values will become another single column, and the name of that variable is set by `values_to`. The output is stored in `df_long`.

See the resulting table in *Figure 8.30*.

	project	months	expenses
1	project1	Jan	1560
2	project1	Feb	1152
3	project1	Mar	1145
4	project2	Jan	1996
5	project2	Feb	1073
6	project2	Mar	1633
7	project3	Jan	1320
8	project3	Feb	1227
9	project3	Mar	1048

Figure 8.30 – The same projects dataset in long format

The inverse transformation, from long to wide, can be used for reasons such as better visualization or for presentations. In the wide format, we lose the duplicates because each month is a column. Let's see how to code that now:

```
# Long to Wide
df_long %>%
  pivot_wider(names_from = 'months',
              values_from = 'expenses')
```

The code is similar to the previous one. We called the `df_long` data frame and piped with the `pivot_wider()` function from `tidyr`. The `names_from` parameter is used to identify the names of the new columns to be created, the month names, and the `values_from` parameter indicates where the values will come from. The result shown in *Figure 8.31* displays the data frame back in the original wide format.

	project	Jan	Feb	Mar
1	project1	1560	1152	1145
2	project2	1996	1073	1633
3	project3	1320	1227	1048

Figure 8.31 – The projects dataset transformed back to a wide format

Reshaping data frames is a good transformation to keep in mind, since the data will not always be ready for analysis, visualization, and modeling and may, therefore, require this type of wrangling.

To complete our journey through the major transformations of data wrangling, let's quickly go over other functions from tidyverse that are worth mentioning.

Do more with tidyverse

Closing this chapter, we will quickly study a few functions from tidyverse that were not mentioned in any of the previous sections but that can be very helpful when solving data wrangling problems.

Consider again the `mtcars` dataset (load it with `data('mtcars')`), which has information about 32 cars from the *1974 Motor Trend Use* magazine. We are already familiar with that data, and we can use it as a reference to learn about the next few transformations.

Let's dive right in on a couple of functions of the `purrr` library. This library brings functions like those from the Apply family, studied in *Chapter 5*. The most interesting function to look at is the `map()` function. It applies the same function to every element of a vector or list. If we want to map the average of the variables' horsepower and weight, this is how to do it:

```
# Map
mtcars %>%
  select(hp, wt) %>%  map(mean)


$hp
[1] 146.6875


$wt
[1] 3.21725
```

In the previous code, we selected the variables for horsepower and weight and requested the calculation of the `mean` for them. It could be any other function, such as `summary`, which would show the descriptive statistic, for example. Observe that `map()` applied the mean calculation to every variable at once, reducing the need to repeat the code. That is the main benefit of functions such as `map`.

There are many kinds of mapping functions that return different types of objects, as well as a few other methods to handle list objects in R. You can see them in the *Further reading* section by navigating to purrr's official documentation page.

From `dplyr`, there are functions used to combine data, such as the binding functions, just like we did when using base R. To bind more rows to your data, given that it has the same columns, we can use `bind_rows()`:

```
# Creating datasets A and B
A <- mtcars[1:3, ]
B <- mtcars[4:6, ]


# Bind rows
AB <- A %>% bind_rows(B)
```

We sliced the first three rows and then rows 4 to 6. Since both subsets come from the same dataset, `mtcars`, we can bind the rows with the `bind_rows()` function, as presented in the previous code snippet.

To bind more columns to the data, we can use `bind_cols()`. The data must have the same number of observations:

```
# Creating datasets A and B
A <- mtcars[1:5, 1:3]
B <- mtcars[1:5, 4:6]


# Bind columns
AB <- A %>% bind_cols(B)
```

Once again, using slices of the same dataset to illustrate the function, if we keep the same observations and slice columns 1 to 3 and 4 to 6, to glue them together is simple if we use `bind_cols()`.

Another interesting set of functions is the cumulative aggregate functions used alongside `mutate()`, such as `cumsum()`, `cumprod()`, `cummean()`, `cummax()`, and `cume_dist()`. These are used for cumulative calculations of a variable using a simple syntax, requiring that we just pass a variable name to it within the `mutate` function, as seen next.

This is the syntax used to calculate cumulative sum of a variable. Let's see the cumulative sum of the weight of the cars in `mtcars`:

```
mtcars %>% mutate( cumulative_weight= cumsum(wt) )
```

The new column is added as the last one on the right-hand side, shown in *Figure 8.32*.

```
                mpg cyl  disp  hp drat    wt  qsec vs am gear carb cumulative_weight
Mazda RX4       21.0   6 160.0 110 3.90 2.620 16.46  0  1    4    4             2.620
Mazda RX4 Wag   21.0   6 160.0 110 3.90 2.875 17.02  0  1    4    4             5.495
Datsun 710      22.8   4 108.0  93 3.85 2.320 18.61  1  1    4    1             7.815
Hornet 4 Drive  21.4   6 258.0 110 3.08 3.215 19.44  1  0    3    1            11.030
Hornet Sportabout 18.7 8 360.0 175 3.15 3.440 17.02  0  0    3    2            14.470
Valiant         18.1   6 225.0 105 2.76 3.460 20.22  1  0    3    1            17.930
```

Figure 8.32 – Cumulative sum of weight from mtcars

The following line of code will calculate the percentage of the distribution for the weights, like a percentile calculation:

```
mtcars %>% mutate(cum_pct= cume_dist(wt) ) %>% arrange(cum_pct)
```

The code will display the next table.

```
               mpg cyl  disp  hp drat    wt  qsec vs am gear carb cum_pct
Lotus Europa   30.4   4  95.1 113 3.77 1.513 16.90  1  1    5    2 0.03125
Honda Civic    30.4   4  75.7  52 4.93 1.615 18.52  1  1    4    2 0.06250
Toyota Corolla 33.9   4  71.1  65 4.22 1.835 19.90  1  1    4    1 0.09375
Fiat X1-9      27.3   4  79.0  66 4.08 1.935 18.90  1  1    4    1 0.12500
Porsche 914-2  26.0   4 120.3  91 4.43 2.140 16.70  0  1    5    2 0.15625
Fiat 128       32.4   4  78.7  66 4.08 2.200 19.47  1  1    4    1 0.18750
```

Figure 8.33 – Cumulative percentage of weight from mtcars

Observe that in *Figure 8.33* the weight variable (wt) was organized before calculating the cumulative percentage. Hence, we can say that 18% of the cars weigh the same or less than 2200 lbs.

Since dplyr was designed with SQL in mind, the library has the popular case_when() function to deal with multiple cases of logical tests. See an example in the following sequence:

```
# Case When to create a new label column for the "am" variable
mtcars %>%
   mutate( transmission_type=
             case_when(
               am == 0 ~ 'automatic',
               am == 1 ~ 'manual' )  )
```

The code calls the mtcars dataset and pipes it with the mutate function, which receives the name of the new transmission_type variable to be created by the case_when() function. This function, by the way, takes the logical tests and the values to be returned if the condition is true, both separated by a ~ tilde mark. If am is 0, return automatic. If am is 1, return manual, as seen in *Figure 8.34*.

```
                  mpg cyl  disp  hp drat    wt  qsec vs am gear carb transmission_type
Mazda RX4         21.0   6 160.0 110 3.90 2.620 16.46  0  1    4    4            manual
Mazda RX4 Wag     21.0   6 160.0 110 3.90 2.875 17.02  0  1    4    4            manual
Datsun 710        22.8   4 108.0  93 3.85 2.320 18.61  1  1    4    1            manual
Hornet 4 Drive    21.4   6 258.0 110 3.08 3.215 19.44  1  0    3    1         automatic
Hornet Sportabout 18.7   8 360.0 175 3.15 3.440 17.02  0  0    3    2         automatic
Valiant           18.1   6 225.0 105 2.76 3.460 20.22  1  0    3    1         automatic
```

Figure 8.34 – Result from the case_when function

Finally, let's briefly mention ggplot2, which is an awesome library. Personally, I think it is one of the best tools for data visualization. We will have a whole chapter dedicated to it later. But since it will be used a few times in the next chapter, we can have a little taste of what's to come. Let's plot a simple graphic using ggplot2.

The basic syntax is as follows. The `ggplot()` function receives the data. Then, we must link the layers using a + plus sign. Next, we must choose our geometry, which is the graphic type (points, bar, histogram, and so on). That geometry receives the `aes()` aesthetics with the x and y axis inside, along with the `color`, `size` of the points, and other arguments to set up the visual elements of the graphic. Generally, it looks like the following code:

```
ggplot( data ) +
    geom_function( aes(x, y, fill) )
```

So, using that explanation, plotting a scatterplot will be like the following code snippet. The `ggplot` function receives the `mtcars` dataset. Then we choose the `geom_point()` geometry to plot a scatterplot. That geometry receives the `aes()` aesthetics with the horsepower (`hp`) on the x axis and miles per gallon (`mpg`) on y axis, along with the `color`, `size` of the points, and the `alpha` parameter, which regulates the opacity of the points. To complete the code, we add a title with `ggtitle()`:

```
# ggplot2 basic scatterplot
ggplot(data= mtcars) +
  geom_point( aes(x=hp, y= mpg),
              color='royalblue', size=4, alpha=0.5 ) +
  ggtitle('Relationship between HP vs. MPG')
```

The plot is returned as seen in *Figure 8.35*.

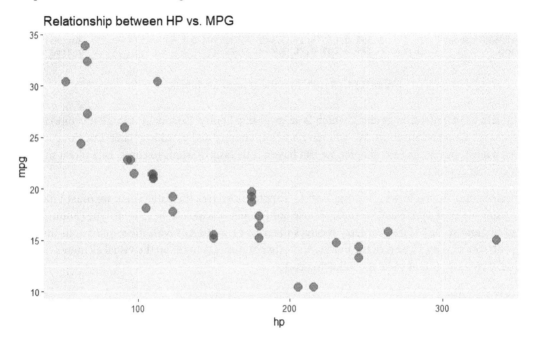

Figure 8.35 – Scatterplot hp versus mpg plotted with ggplot2

We close this chapter with that brief introduction to a plot with ggplot2.

In the final chapter of *Part 2*, we will perform a complete exploratory data analysis in R, gathering many of the data wrangling concepts learned up till now, thus increasing our knowledge and experiencing of how an EDA can be done in a practical project.

Summary

We have been through a lot in this chapter. After all, the core of data wrangling is about the transformations that we covered here. Most of the work of analysts, data scientists, and developers involves filtering, reshaping, joining, and doing all kinds of data wrangling to get the data into the shape needed for the project.

We started by studying slicing and filtering, allowing us to zoom in on parts of a dataset and revisited the difference between both. Then, we moved on to show you how to group and summarize data, which turns out to be an important task, possibly one of the most used functions when munging data.

Replacing and filtering were the next two subjects. That section covered how to replace values and how to fill in missing data in our dataset. Once the data was cleaned of *NAs*, the subsequent step was ordering the data, making it easier to read and understand the information.

It is worth repeating that datasets have more information than what it may initially look like. If we split, combine, or create custom variables, good data can come out. We learned how to do that and how much the `mutate()` function helps in that process.

Moving closer to the end, we covered the different ways to join datasets, given that data can be stored in different sources or tables. We also learned how to reshape data, transforming it to long or wide formats, and closed the chapter with a handful of functions to keep as potential helpers for data wrangling activity.

The next chapter is the closure of *Part 2* of this book. We are ready to work on a complete EDA practical project now, using the vast knowledge built so far.

Exercises

1. What function is used to filter data?
2. What happens when `group_by()` is used without the summarizing function?
3. Write the code to return the `capital_gain` mean grouped by `native_country`.
4. What's function is used to drop all *NAs* from a dataset?
5. Write the code to replace *NAs* with another value.
6. Write the code to return the average net gain (`capital_gain - capital_loss`) by education level, arranged in decreasing order.
7. What is the main function to be used to create new variables?
8. List all the types of joins.
9. What is the difference between left join and inner join?
10. What are the two functions used to reshape data?

Further reading

- Dplyr official documentation and cheat sheet: `https://dplyr.tidyverse.org/`

- Tidyr official documentation and cheat sheet: `https://tidyr.tidyverse.org/`

- Readr official documentation and cheat sheet: `https://readr.tidyverse.org/`

- Purrr official documentation and cheat sheet: `https://purrr.tidyverse.org/`

- The power of mutate for data wrangling: `https://tinyurl.com/y8fdjpdv`

- SQL reference documentation from Oracle: `https://docs.oracle.com/cd/B19306_01/server.102/b14200/toc.htm`

- Types of joins:

  `https://tinyurl.com/yexdrev5`

  `https://en.wikipedia.org/wiki/Join_(SQL)`

- Code for this chapter in GitHub: `https://tinyurl.com/yckswc37`

9

Exploratory Data Analysis

Exploratory data analysis, also referred to as **EDA**, is as important as the other steps in a Data Science project. It helps one to deeply understand the data and capture deviances that can harm the modeling. After all, we know that *garbage in* will result in *garbage out*.

There are some steps used to perform data exploration, and that is what we will cover in this chapter. The intent is to go over a practical project, beginning with the dataset load to RStudio until the composition of an EDA report, outlining the most interesting findings. The steps to be covered in this chapter are as follows:

- Loading the dataset to RStudio
- Understanding the data
- Treating missing data
- Exploring and visualizing the data
- Analysis report

Technical requirements

The dataset chosen for this project is from *American Community Survey 2010-2012 Public Use Microdata Series* and can be found in the raw form at this link: `https://tinyurl.com/4yx2dhpe`. The data is originally from the US census, but *FiveThirtyEight* pulled it and did some cleanup, creating the final dataset to be used in this chapter. The links to the raw data can also be found in the previously provided URL address. The data is about college majors and there is a lot of information about earnings, employment ratio, and share of enrolled people by sex, thus containing many numbers that we can use to understand how the majors and areas are performing in the job market and in terms of unemployment rates.

All the code can be found in the book's GitHub repository: `https://github.com/PacktPublishing/Data-Wrangling-with-R/tree/main/Part2/Chapter9`.

Load the following libraries for this chapter:

```
library(tidyverse)
library(skimr)
library(statsr)
library(GGally)
library(corrplot)
```

Let's move on and start the EDA in practice.

Loading the dataset to RStudio

First, we need to load the libraries to be used in this EDA. We are going to need the libraries loaded as follows. As seen in *Chapter 8*, the `tidyverse` package is composed of eight core libraries, being a robust tool to work with data in R. `skimr` will be useful for the descriptive statistics calculations and display, and `statsr` is a great library that brings us many statistical tools, which we will be using to help with data sampling, more specifically. `GGally` is used for pair plots and `corrplot` for correlation plots:

```
library(tidyverse)
library(skimr)
library(statsr)
library(GGally)
library(corrplot)
```

To load the dataset to RStudio, we can use the `read_csv()` function. As we have seen many times so far, that function is able to read CSV files directly from a web address, so that is what we will do in the subsequent code. We define a `url` variable with the website address and add that to the reading function:

```
# Path
url <- "https://raw.githubusercontent.com/fivethirtyeight/data/
master/college-majors/recent-grads.csv"

# Load dataset to RStudio
df <- read_csv(url)
```

Concerning dataset load, that is pretty much what is to be done. We can add another step if we want to save a copy of the dataset to another variable just in case we do any transformation that needs to be undone or if we need to revisit the original form of the data for comparison or recovery:

```
# Keep a copy of our original data
df_original <- df
```

With that done, now it is time to take our first look at the dataset and start understanding it.

Understanding the data

Once the data is up in RStudio, we should look at it. The first checkup points are to confirm that the data was fully loaded and without errors. For that, we can use the software's built-in viewer by typing `View(df)`, or we can use the `head()` function to look at just a couple of lines, remembering that the default is to show the first six observations:

```
df %>% head()
```

It displays the following result:

A tibble: **6 x 21**

Rank <dbl>	Major_code <fctr>	Major <fctr>	Total <dbl>	Men <dbl>	Wom... <dbl>
1	2419	PETROLEUM ENGINEERING	2339	2057	282
2	2416	MINING AND MINERAL ENGINEERING	756	679	77
3	2415	METALLURGICAL ENGINEERING	856	725	131
4	2417	NAVAL ARCHITECTURE AND MARINE ENGINEERING	1258	1123	135
5	2405	CHEMICAL ENGINEERING	32260	21239	11021
6	2418	NUCLEAR ENGINEERING	2573	2200	373

6 rows | 1-6 of 21 columns

Figure 9.1 – Head of the College Majors dataset

After a first look, the data looks good. In general, what we are looking for here are the following:

- Whether the data is **rectangular**—in other words, divided into rows and columns.

- Whether we see any problems with language encoding, which, when it occurs, shows some symbols amidst the words.

- Whether the CSV reading was successful for all columns because if the separator for the file is a semicolon or tab, for example, the columns can appear all merged together.

Since there are none of those problems, let's move on.

The next step is to call the `glimpse(df)` function, which will check the data types of each variable, as well as inform about the number of rows and columns. *Figure 9.2* shows the result:

```
Rows: 173
Columns: 21
$ Rank                <dbl> 1, 2, 3, 4, 5, 6, 7, 8, 9, 10, 11, 12, 13, 14, 15, 16, 1~
$ Major_code          <dbl> 2419, 2416, 2415, 2417, 2405, 2418, 6202, 5001, 2414, 24~
$ Major               <chr> "PETROLEUM ENGINEERING", "MINING AND MINERAL ENGINEERING~
$ Total               <dbl> 2339, 756, 856, 1258, 32260, 2573, 3777, 1792, 91227, 81~
$ Men                 <dbl> 2057, 679, 725, 1123, 21239, 2200, 2110, 832, 80320, 655~
$ Women               <dbl> 282, 77, 131, 135, 11021, 373, 1667, 960, 10907, 16016, ~
$ Major_category      <chr> "Engineering", "Engineering", "Engineering", "Engineerin~
$ ShareWomen          <dbl> 0.12056434, 0.10185185, 0.15303738, 0.10731320, 0.341630~
$ Sample_size         <dbl> 36, 7, 3, 16, 289, 17, 51, 10, 1029, 631, 399, 147, 79, ~
$ Employed            <dbl> 1976, 640, 648, 758, 25694, 1857, 2912, 1526, 76442, 619~
$ Full_time           <dbl> 1849, 556, 558, 1069, 23170, 2038, 2924, 1085, 71298, 55~
$ Part_time           <dbl> 270, 170, 133, 150, 5180, 264, 296, 553, 13101, 12695, 5~
$ Full_time_year_round <dbl> 1207, 388, 340, 692, 16697, 1449, 2482, 827, 54639, 4141~
$ Unemployed          <dbl> 37, 85, 16, 40, 1672, 400, 308, 33, 4650, 3895, 2275, 79~
$ Unemployment_rate   <dbl> 0.018380527, 0.117241379, 0.024096386, 0.050125313, 0.06~
$ Median              <dbl> 110000, 75000, 73000, 70000, 65000, 65000, 62000, 62000,~
$ P25th               <dbl> 95000, 55000, 50000, 43000, 50000, 50000, 53000, 31500, ~
$ P75th               <dbl> 125000, 90000, 105000, 80000, 75000, 102000, 72000, 1090~
$ College_jobs        <dbl> 1534, 350, 456, 529, 18314, 1142, 1768, 972, 52844, 4582~
$ Non_college_jobs    <dbl> 364, 257, 176, 102, 4440, 657, 314, 500, 16384, 10874, 5~
$ Low_wage_jobs       <dbl> 193, 50, 0, 0, 972, 244, 259, 220, 3253, 3170, 980, 372,~
```

Figure 9.2 – Result of glimpse (df) function

The data has 173 rows (observations) and 21 columns (variables). Almost all the variables are numeric, of the `double` type. `Major` and `Major_category` are of the `character` type. These last two are categories, thus they must be transformed into the `factor` type. `Major` and `Major_code` are related to each other, but even though `Major_code` is numeric, it is actually a factor as well. So, we will transform those three variables to `factor`, as follows, using the `mutate_at()` function, which takes in the columns to be transformed and the data type to convert to:

```
# Columns to change to factor
cols_to_factor <- c("Major", "Major_code", "Major_category")


# Assign variables as factor
df <- df %>%
  mutate_at(cols_to_factor, factor)
```

If we call `glimpse(df)` again, it will show the variables as the factor (`<fct>`) type now.

As part of understanding the data, looking at descriptive statistics is imperative since it can bring up many insights, especially those related to the distribution of the variables. Here, we use central location measurements to help us understand the data. Next, we will use the `skim()` function to look at the descriptive stats of the *College Majors* dataset. First, this subsequent code changes the option of displaying scientific notation for the numbers to show them in thousands and three decimal digits:

```
# Remove scientific notation
options(scipen=999, digits = 3)
```

Then, we will manually calculate the **coefficient of variation (CV)** for the variables, as this information is not available in skimr's summary. To calculate that, we will run the skim() function and store the resulting table in a variable called stats. From that table, we get the standard deviation variable values and divide them by the mean variable values:

```
# Coefficient of Variance
stats <- skim(df)
stats$numeric.sd/stats$numeric.mean
```

This is the output:

```
# OUTPUT:
[1]    NA    NA    NA 0.576 1.612 1.682 1.813 0.443 1.737 1.625
1.647 1.658 1.684
[14] 1.702 0.445 0.286 0.311 0.289 1.729 1.791 1.800
```

The NA values in this output are referent to the three factor variables since we can't divide non-numeric elements.

Finally, let's run the skim() function once more to see the descriptive statistics:

```
# Descriptive stats
skim(df)
```

The resultant summary table is very complete, and that is why it is easier to use this library instead of wasting time typing code to do all of this. *Figure 9.3* shows a partial screenshot of the table. In addition to what you see in the figure, there are also the minimum, 25th percentile, median, 75th percentile, and maximum values:

A tibble: 18 x 11

	skim_variable	n_missing	complete_rate	mean	sd
	<chr>	<int>	<dbl>	<dbl>	<dbl>
1	Rank	0	1.000	87.0000	50.0849
2	Total	1	0.994	39370.0814	63483.4910
3	Men	1	0.994	16723.4070	28122.4335
4	Women	1	0.994	22646.6744	41057.3307
5	ShareWomen	1	0.994	0.5222	0.2312
6	Sample_size	0	1.000	356.0809	618.3610

Figure 9.3 – Partial view of the summary table with descriptive statistics from skimr

From the result, shown partially in the table in *Figure 9.3*, the following insights were extracted:

- We have 173 observations and 21 variables (18 numeric and 3 factor).

- There is only one observation with missing data present on Men, Women, ShareWomen, and Total. This is probably the same observation.

- On average, there are more women enrolled than men overall:

  - Around 52% of those enrolled in the majors considered are women.

- There are many more people employed (~31k) than unemployed (~2.5k), which aligns with the 6% unemployment rate average.

- Salaries are somewhere between $30k and $52k a year, on average.

- There is a balance between jobs requiring a college degree versus not requiring it.

- Looking at the coefficients of variance calculated, we see that the data is very spread or with big tails, which we will confirm with the distribution visualizations.

Now, we already have a high-level understanding of the data. We know how many observations and variables there are in it, as well as the data types—those will be important to know which functions can be used with each variable, as well as the best kind of graphic to build with them. We also have a good idea of how spread the data points are after we looked at the descriptive statistics.

The next section covers how to treat missing data.

Treating missing data

Usually, those observations won't be valid for statistics calculations, as there is no value present. Therefore, despite having calculated the descriptive statistics before handling the missing values, it won't affect our results or insights. However, for the continuation of the data analysis, we must handle the NA values to understand whether those carry a meaning or not and then decide how to proceed with them.

A missing data point is also information. It can mean that the data was erroneously missed by human or system error, or it can mean that a person did not respond to a question, for example. So, if we were dealing with a system log and seeing a bunch of NA values, it would be necessary to check whether the measurements were being correctly registered or whether those missing data points should be expected. Another example to be considered: on poll data, if there are a lot of missing answers, it can be either that nobody is answering the question or that the person chose to answer *N/A*, which should be assessed and documented because *N/A*, in this case, is an option, not a missing value. Consequently, these nuances and peculiarities must be accounted for before we just remove the missing values or input something else for them.

Back to our project, the missing value, in this case, is on *line 22*, shown in *Figure 9.4*:

	Rank	Major_code	Major		Total	Men	Women	M
22	22	1104	FOOD SCIENCE		*NA*	*NA*	*NA*	Ac

Figure 9.4 – Missing value in the College Majors dataset

There is no data for the `Total`, `Men`, `Women`, and `ShareWomen` variables for that observation. Since it is only 1 observation out of 173, representing less than 1% of the data, we can just drop that major. It won't affect our result and it will be more accurate than if we input another number in there, such as the average or median value.

To do that, the code is simple. Just use the `drop_na()` function from `tidyr`:

```
# Drop NA
df_clean <- df %>% drop_na()
```

Now, the dataset has 172 observations, as displayed by the `dim()` function, which pulls the dimensions of the dataset:

```
# New Dimensions
dim(df_clean)
[1] 172  21
```

It is time to move forward and start exploring the data. To do that, we will use a questions-and-answers approach, using data and plotting some visuals to answer our exploratory questions.

Exploring and visualizing the data

The exploration phase of an EDA is the main portion of it, naturally. In this section, the idea is to take a thorough look at the variables, understand their distributions, start creating some questions that will lead the exploration, and use the data to answer them.

Univariate analysis

The first step to take concerns univariate analysis—looking at one variable at a time. The best approach is to create some histograms to look at the distribution of the variables. According to Hair Jr. et al. (2019), plotting the variables' distributions and looking at their shape is a good point to start understanding the nature of those variables.

In the next code snippet, we will loop through all the numeric variables and plot one histogram for each. It starts with a `for` loop to iterate over each variable in the column names that presents the numeric type (there is the importance of knowing the data types, from previous sections).

If it is numeric, then it creates a plot with the basic function to create plots, ggplot(), adding the histogram geometry with the geom_histogram() function, which receives the aes() mapping aesthetics with the values from that variable. To give only the values to the aesthetics, we must use the unlist() function. The complement of the code is just some plot settings such as bins=20, fill color (fill="royalblue"), and border color (color= "gray"), as well as the title, defined by ggtitle(), and the x label, defined by labs(). This entire graphic is stored in a variable named g and plotted using the plot(g) command for each variable:

```
# Loop through numeric columns
for ( variable in colnames(select_if(df_clean, is.numeric)) ) {
  g=ggplot(df_clean) +
    geom_histogram(aes(unlist(df_clean[, variable]) ), bins=20,
                   fill="royalblue",    color="gray")+
    ggtitle( paste("Histogram of", variable) ) +
    labs(x= variable)
  plot(g)
}
```

As a result, there will be 18 histograms created. In *Figure 9.5*, you can see four of them, in order: **Histogram of Employed**, **Histogram of Full_time**, **Histogram of Unemployment_rate**, and **Histogram of Median**:

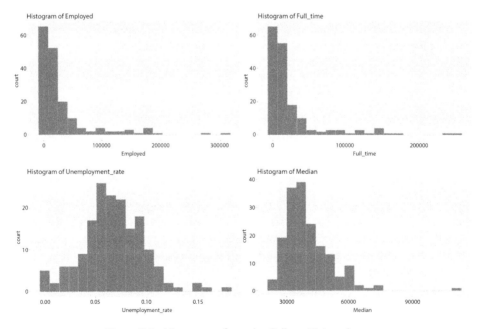

Figure 9.5 – Histograms from the College Majors dataset

The histograms show some heavily right-skewed variables in the top two charts, while the bottom charts are slightly closer to a normal distribution, but still present many possible outliers. To be able to confirm the presence of values over the acceptable range of variance, a good approach is to plot boxplots, a type of graphic that already brings the outlier values calculated using the quartiles' values Q1 and Q3 added by 1.5 times the difference between these two values, as discussed in *Chapter 3*.

Let's do just what we did with the histograms and plot a boxplot for each numeric variable:

```
# Loop through numeric columns
for ( variable in colnames(select_if(df_clean, is.numeric)) ) {
  g=ggplot(df_clean) +
    geom_boxplot(aes(y=unlist(df_clean[, variable]) ),
                 fill="royalblue", color="gray") +
    ggtitle( paste("Boxplot of", variable) ) +
    labs(y= variable)
  plot(g)
}
```

We should not spend too much time describing the preceding code snippet because it is basically the same as the one used to plot the histograms, with the difference being that the geometry used in this case was `geom_boxplot()`.

Once again, four of the resulting graphics are displayed in the sequence, for the same variables as the previous histogram plots:

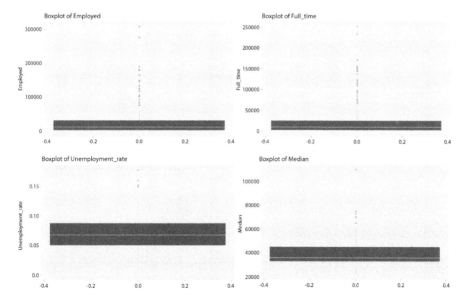

Figure 9.6 – Boxplots of the College Majors dataset

Looking at *Figure 9.6*, there are clearly many outliers, represented by circles over the superior whisker of the box. Observe that the top plots have many more outliers than the bottom ones since the last ones closely resemble a normal distribution, thus the majority of their values will be found closer to the mean.

The visuals presented before are enough for this univariate analysis. We could maybe add QQ plots as well, and this would confirm whether any of these variables are normally distributed. However, in this case, this conclusion can be clearly made just by looking at the histograms and boxplots. The histograms show highly skewed data and the boxplots have many outliers popping out of the box. This is sufficient to point us in the right direction. It is recommended, though, for data looking more like a bell-shaped curve, that you plot the QQ plot to have a visual indication of normality. Another option, and maybe quicker in terms of coding, is to run a normality test for the variables in question.

In this dataset, the variable that more looks like a normal distribution is `Unemployment_rate`, printed next, in *Figure 9.7*.

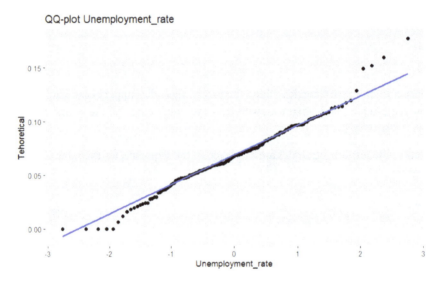

Figure 9.7 – QQ plot of Unemployment_rate

Notice that the deviances occur in the tails, but most of the data is very close to normality. Running the **Shapiro-Wilk** test of normality using `shapiro.test(df_clean$Unemployment_rate)`, and considering 5% of the significance level, let's state the hypothesis:

- *Null hypothesis (p-value > 5%)*: The data comes from a normal distribution

- *Alternative hypothesis (p-value <= 5%)*: The data does not come from a normal distribution

If we run that test in RStudio, the result is *p-value = 0.02*. Since the p-value is under the significance level of 5%, there is statistical evidence to reject the null hypothesis in favor of the alternative hypothesis, leading us to conclude that the data does not come from a normal distribution.

Given that we now have a deeper understanding of the variables' distributions, we will move forward to multivariate analysis.

Multivariate analysis

Now, it is time to check how the variables relate to each other. In a project, the idea is to explore how the explanatory variables (X) affect the response variable (y), which helps determine the best ones for a posterior model.

In this project, we are interested in seeing how the variables affect the `Unemployment_rate` variable. To start this exploration, the use of bivariate plots such as scatterplots is ideal for numeric features, such as the ones in this project. A scatterplot is the most popular method of examining bivariate relationships (Hair et al., 2019). This graphic type provides a quick view of the pattern created when two variables are compared, becoming a great ally to determine whether there is correlation—a statistical test to determine the strength of the linear relationship between two variables. The more correlated they are, the more the scatterplot will look like a line.

A recommended plot that gathers both presentations is a pair plot, which brings the scatterplots on the lower diagonal and the correlation test numbers on the upper diagonal. The center part shows the density distribution of the variables or the histograms, depending on the library chosen.

To create a pair-plot matrix for the entire data, we can use the `ggpairs()` function from `GGally`. Follow the code shown next:

```
# Check linear relationship and correlations
ggpairs(df_clean[, -c(1,2,3,7)])
```

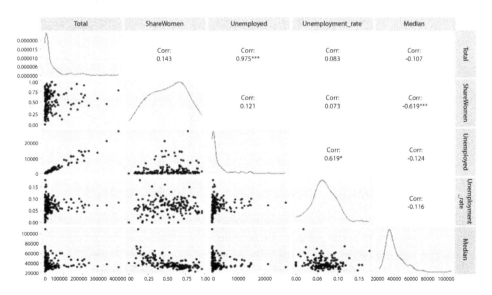

Figure 9.8 – Pair plot showing linear relationships and correlations

In *Figure 9.8*, there is only an extract of part of the pair plot for the *College Majors* dataset. The entire matrix, in this case, is too big to be printed. Observe that the correlations between other variables and `Unemployment_rate` are weak, staying under 10% to 20% (`Total`: 0.083, `ShareWomen`: 0.073, `Unemployed`: 0.169, `Median`: -0.116). There are other relationships with `Unemployed` that are strong, though, such as its correlation with `Total`—around 97%. This means that the variables are able to explain better the variance of `Unemployed` than `Unemployment_rate`, a possible problem for modeling.

An alternative to a pair plot is a correlation matrix. It does not bring scatterplots with it, but it is also very useful. The `corrplot` library has a homonymous function to draw a customizable matrix with all the correlation numbers, providing an interesting view of the correlations.

The code to create such a plot is presented in the following sequence, where a rounded correlation matrix is created with `cor()`, excluding the factor variables and the rank numbers. That is assigned to `correlations`. Next, we provide `correlations` to the `corrplot()` function, specifying some visual configurations, such as `method= "number"` to see numbers (there are other options available) and `type= "lower"` to show only the lower portion of the plot and avoid redundancy, while `tl.cex` and `number.cex` are the sizes of the labels' text and numbers inside the matrix, respectively:

```
# Correlations plot between numeric variables
correlations <- round(cor(df_clean[, -c(1,2,3,7)]),2)
corrplot(correlations, method="number", type="lower",
         tl.cex = 0.8, number.cex = 0.6)
```

This code displays the following output:

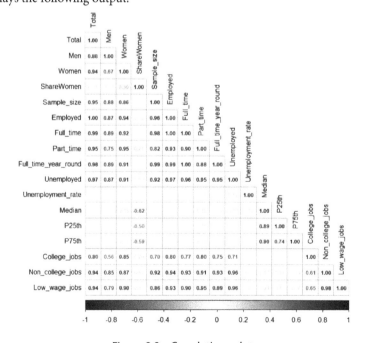

Figure 9.9 – Correlations plot

In *Figure 9.9*, the closer the correlation is to zero, the lighter the number is. Once again, it is clear that the correlations between `Unemployment_rate` and the other variables are very weak. If the intention is to model that rate using linear regression, for example, the results are not very promising so far. Given that the linear regression is the square of the correlation value, the R^2 in this case would probably not be very high without any changes to the current data.

Another point to be observed is the high quantity of multicollinear variables. Multicollinearity is when two explanatory variables have a strong linear relationship between them. In that case, both will be capable of explaining the same variance in the target variable, creating redundancy and making it more difficult to determine the effect of each individual explanatory feature on the target. That effect generates unstable regressions (Bruce & Bruce, 2019); therefore, it must be eliminated before modeling. As we are just exploring the data, that task will not be performed.

Once the bivariate initial analysis is completed, the continuation can be done by formulating questions to lead the exploration.

Exploring

Let's start formulating questions that will guide the sequence of the EDA. The first question to respond to is this: *What are the top 10 majors with the lowest unemployment rate?* Take a look at the following code snippet:

```
# Select only top 10 for plotting
top10_low_unemploy <- df_clean %>%
  select(Major, Unemployment_rate) %>%
  arrange(Unemployment_rate) %>%
  head(10)
```

To begin answering the first exploratory question, in the preceding code snippet we created a new subset by selecting the majors and unemployment rates, arranging them by `Unemployment_rate`, and taking only the top 10 observations with `head(10)`. That will be assigned to `top10_low_unemploy`.

In the sequence, we use that variable just created to generate a graphic using `ggplot`. The plotting function receives the `top10_low_unemploy` variable, then we add the `geom_col()` geometry providing the x and y axes, border color, and `fill`. The label on the y axis is set up with `labs()`, and we are adding value labels with `geom_text()` and a title with `ggtitle()`:

```
# plot
ggplot(top10_low_unemploy) +
  geom_col( aes(x=Unemployment_rate,
               y=reorder(Major, -Unemployment_rate) ),
           color="royalblue", fill= "royalblue") +
```

```
labs(y= "Major") +
geom_text(aes(x = Unemployment_rate,
              y= Major, label = round(Unemployment_rate,3)),
          size=3, hjust = 1) +
ggtitle("Unemployment Rate by Major")
```

The code plots the following output:

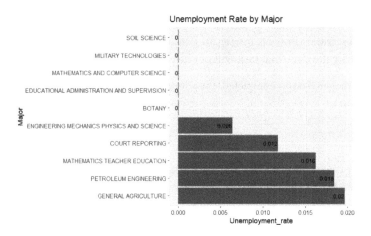

Figure 9.10 – Top 10 majors with lower unemployment rates

Figure 9.10 shows the top 10 college majors with lower unemployment rates; however, five of those are zeros. If we look at the data, we will see that those majors have just a small number of people enrolled, most times below 100, thus the numbers won't be representative of reality. Think about it: if a sample of 100 has 99 people employed, it is 99%, while the same number out of a sample in the thousands mark means almost nothing. In that case, the solution is to build the same visual using a proportional rate of people enrolled and check the unemployment rates again.

The code to create the proportional subset is shown subsequently, where df_clean has an added feature named proportion, created with the mutate() function. Then, we select the Major, Unemployment_rate, and proportion variables, arrange proportion by descending order and Unemployment_rate in ascending, and finally filter only the top 10 results:

```
# Select only top 10 for plotting
top10_proportional <- df_clean %>%
  mutate(proportion = Total/ sum(Total) ) %>%
  select(Major, Unemployment_rate, proportion) %>%
  arrange(desc(proportion), Unemployment_rate ) %>%
  head(10)
```

The code for the plot is pretty much the same as for the output in *Figure 9.10*, just changing the subset name to `top10_proportional` and changing the title. The result is shown in *Figure 9.11*:

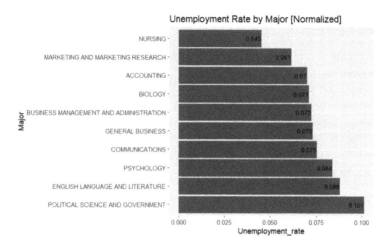

Figure 9.11 – Normalized top 10 majors with lower unemployment rates

The plot in *Figure 9.11* looks more interesting now. Of the more popular majors, proportionally speaking, nursing is the one with the lowest unemployment rate, with only 4.5% of the students not employed.

Looking for the areas that are more specialized, meaning that there are more jobs requiring college, *what are the majors with more jobs requiring college (more specialized)?*

We can, initially, compute the percentage of college jobs over the total for each major and add that as a new variable, using the `mutate()` function:

```
# Difference collge - non-college jobs
df_clean <- df_clean %>%
  mutate( College_jobs_pct = College_jobs/ (College_jobs + Non_
college_jobs)  )
```

Then, if we group the data by major category and calculate the mean of that percentage, ordering the values in descending order by the mean, here is what the code looks like:

```
# More specialized by major category
df_clean %>%
  select(Major_category, Major, College_jobs_pct) %>%
  group_by(Major_category) %>%
  summarise(mean_coll_pct= mean(College_jobs_pct)) %>%
  arrange( desc(mean_coll_pct) )
```

The resulting output is displayed next:

A tibble: 16 x 2

Major_category	mean_coll_pct
<chr>	<dbl>
Education	0.714
Engineering	0.680
Computers & Mathematics	0.604
Biology & Life Science	0.593
Interdisciplinary	0.570
Physical Sciences	0.532
Health	0.516
Psychology & Social Work	0.508
Agriculture & Natural Resources	0.406
Humanities & Liberal Arts	0.386
Social Science	0.375
Communications & Journalism	0.348
Arts	0.328
Law & Public Policy	0.324
Business	0.297
Industrial Arts & Consumer Services	NaN

Figure 9.12 – Major categories with more college-required jobs

Figure 9.12 presents the education area at the top, with over 70% of the jobs requiring college, followed by **Science, Technology, Engineering, and Mathematics (STEM)**-related majors. The bottom position disregarding the NA value is for business-related majors.

The next question we are interested in answering is this: *What are the best median-value-paying jobs?*

The approach is to create a boxplot, enabling the visualization of the median values, as well as the variability of each major, on a single graphic.

Here, we will begin adding a median_pay variable to the dataset, using group_by() to group the categories and mutate() to calculate the median pay:

```
# Add variable median salaries
df_clean <- df_clean %>%
  group_by(Major_category) %>%
  mutate(median_pay= median(Median))
```

Then, to plot the graphic, the code snippet is explained in the following sequence.

Using the basic `ggplot()` plotting function and providing it with the dataset, we add the geometry boxplot. The x axis will be arranged by `reorder()` of the majors based on the new `median_pay` variable. The setups are similar to what we have seen before, such as filling color, `labs()`, and `ggtitle()`. New here are the `theme()` function, which adds an angle of 45 degrees to the x ticks, `expand_limits()`, used to enlarge the space for the ticks, allowing it to show thousands, and `scale_y_continuous()`, to add a comma to the y ticks:

```
# Boxplots of the median payments by major
ggplot(data= df_clean) +
  geom_boxplot( aes(x=reorder(Major_category, median_pay),
y=Median ),
             fill="royalblue") +
  labs(x= "Major_category") +
  ggtitle("Median by Major Category") +
  theme(axis.text.x=element_text(angle=45, hjust=1)) +
  expand_limits(x = c(0, NA), y = c(0, NA)) +
  scale_y_continuous(labels = scales::comma)
```

The code will result in the following output:

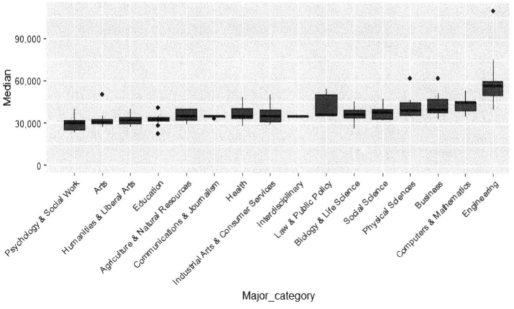

Figure 9.13 – Boxplots of median salaries for each major category

Figure 9.13 shows that the STEM area is among the top-paying jobs. This screenshot is easy to digest and provides similar information to what is received from statistical tests such as ANOVA and t-tests (Hair et al., 2019). Furthermore, the size of the boxes will give an idea of how much variance there is in each group. Look at **Law & Public Policy**, for example, which has a high variance, but the median is almost attached to the bottom of the box.

Moving on with this EDA, the next question that pops up is this: *Do the majors with more share of women enrolled have a higher or lower unemployed rate?*

A simple correlation test can show the answer. Since the data is not normally distributed, the Spearman method is more suited for a correlation test here:

```
# Correlation between ShareWomen vs Unemployment_rate
cor(df_clean$ShareWomen, df_clean$Unemployment_rate,
    method = "spearman")
OUTPUT:
[1]  0.0663
```

The result is close to zero, which means a weak linear relationship between the variables. But let us plot a scatterplot to analyze whether there is any kind of association.

The code snippet that follows is nothing new—a plotting function, adding points geometry with x and y axes and color. The difference is that we are using the `labs()` function to add a title and a subtitle, as well as the `theme()` function to configure the color and size of the subtitle:

```
ggplot( data= df_clean ) +
  geom_point( aes(x=ShareWomen, y=Unemployment_rate),
                color="royalblue") +
  labs(title="Share of women enrolled vs. Unemployment rate") +
  labs(subtitle = "There is no linear relationship between the
two variables, thus the graphic is spread on x and y axes.",
color="darkgray", size=8) +
  theme(plot.subtitle = element_text(color = "darkgray",
size=10) )
```

The scatterplot display looks like this:

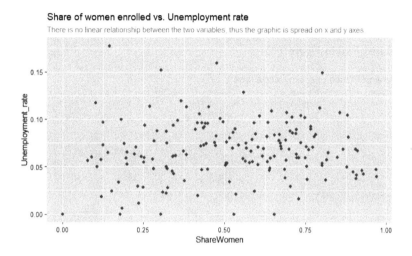

Figure 9.14 – Scatterplot of share of women versus the unemployment rate

As observed in *Figure 9.14*, there is no pattern in the scatterplot. That was expected since the correlation value was close to zero.

Now, the question to follow is regarding the share of women enrolled in the majors and the quantity of low-wage jobs. *Do the majors with a greater share of women enrolled have a similar salary median?*

Once again, if we do a correlation test, the answer will be presented. Then, by plotting a scatterplot using similar code from the previous question, here is what it will look like:

```
# Correlation between the variables
cor(df_clean$ShareWomen, df_clean$Low_wage_jobs)
[1] 0.1878496


# Plot ShareWomen vs Low Wage
ggplot(data=df_clean) +
  geom_point( aes(x=ShareWomen, y= Low_wage_jobs),
              color="royalblue")+
  ggtitle("Share of Women vs. Low wage jobs") +
  labs(subtitle= "The relationship between the variables is
still weak [~19%], suggesting that the share of women enrolled
in
  a major does not affect the wages for the major category." )
```

```
+
  theme(plot.subtitle = element_text(color = "darkgray",
size=10) )
```

The output plotted is shown in *Figure 9.15*:

Figure 9.15 – Scatterplot of share of women enrolled against low-wage jobs

Most of the points are following the x axis without going up too much.

We can go a little further and verify whether the mean values of salaries from majors with more than 50% women are different than the majors with less than 50% women. We will do that using a hypothesis test. First, we need to create two groups with summarized data for majors with more or less than a 50% share of women. The code uses mutate() in combination with the ifelse() function to make a logical test and return a value accordingly, and then add the new over_under variable:

```
# Create new column for Share Women higher or lower than 50%
df_clean <- df_clean %>%
  mutate( over_under=
          ifelse(ShareWomen > 0.5, "higher", "lower") )
```

Second, we must separate it into two sets. The task is completed with the filter and select functions:

```
# Higher than 50% women
higher_women <- df_clean %>%
  filter(over_under == "higher") %>%
```

```
  select(Major, Major_category, Low_wage_jobs)

# Lower than 50% women
lower_women <- df_clean %>%
  filter(over_under == "lower") %>%
  select(Major, Major_category, Low_wage_jobs)
```

Finally, we can test the normality of the subsets using the Shapiro-Wilk test.

Hypothesis test

- *Significance level*: 0.05

- *Ho (p-value >= 0.05)*: The dataset follows a normal distribution

- *Ha (p-value < 0.05)*: The dataset does not follow a normal distribution

```
# Normality tests
shapiro.test(higher_women$Low_wage_jobs)
shapiro.test(lower_women$Low_wage_jobs)


  Shapiro-Wilk normality test
data:  higher_women$Low_wage_jobs
W = 0.6, p-value = 0.00000000000001


  Shapiro-Wilk normality test
data:  lower_women$Low_wage_jobs
W = 0.5, p-value = 0.00000000000001
```

Both distributions are not normal, as we saw in the normality tests. But to perform a statistical t-test, one of the requisites is to have close-to-normal distributions. Thus, before we move on with the t-test to check the differences between the averages of both groups, let's create a sampling of these two datasets to create close-to-normal datasets of averages.

The code in the following sequence is using the `rep_sample_n()` function from the `statsr` library and is taking 100 samples of 1,000 observations, allowing repetitions, and then taking the mean of each sample. This dataset of the means of the samples will be close to normal, as per the **central limit theorem**, or **CLT** (https://tinyurl.com/347cxdwc):

```
# Sampling from Higher Share Women
higher_women_n <- higher_women %>%
  rep_sample_n(size = 1000, reps = 100, replace = TRUE) %>%
```

```
    summarise(mu = mean(Low_wage_jobs))


# Sampling from Lower Share Women
lower_women_n <- lower_women %>%
  rep_sample_n(size = 1000, reps = 100, replace = TRUE) %>%
  summarise(mu = mean(Low_wage_jobs))
```

Now, we are ready for the t-test. Here is the hypothesis:

- *Significance level*: 0.05

- *Ho (p-value >= 0.05)*: Neither average is statistically different

- *Ha (p-value < 0.05)*: Both averages are statistically different

```
# T-test to check if the averages of both groups
t.test(x=higher_women_n$mu, y=lower_women_n$mu,
       alternative = "two.sided")


  Welch Two Sample t-test

data:  higher_women_n$mu and lower_women_n$mu
t = 60, df = 192, p-value <0.0000000000000002
alternative hypothesis: true difference in means is not equal
to 0
95 percent confidence interval:
 1821 1945
sample estimates:
mean of x mean of y
     4742      2859
```

The p-value is close to zero, indicating that we can reject the null hypothesis in favor of the alternative. There is statistical evidence indicating that the averages are different. In practice, we can infer with 95% confidence that the majors with 50% or more of those enrolled being women have, on average, more low-wage jobs related to them.

A visual result of that can be provided by boxplots, as stated before. This code is grouping the data by over_under and creates a new med_sal variable with the mean salary calculation:

```
# Group data
median_salary_w <- df_clean %>%
  group_by(over_under) %>%
  summarise(med_sal= mean(Median))
```

Next, to create the boxplots, the code is like that for the previous blocks:

```
# Plot a Boxplot graphic
ggplot( df_clean ) +
  geom_boxplot( aes(x=over_under, y=Median), fill="royalblue" )
+
  ggtitle("Average Salary When ShareWomen is lower/higher than
50%") +
  labs(subtitle = "The average salary for majors with more
women enrolled is lower than the majors with less women,
  reinforcing the perception that women are getting lower
salaries.",
        x= "50% Share of Women", y= "Mean Salary" ) +
  geom_text(data=median_salary_w, aes(x = over_under, y= med_
sal, label = round(med_sal)),
            size=2.2, vjust = 1, color="white") +
  theme(plot.subtitle = element_text(color = "darkgray",
size=10) )
```

It will result in the following output:

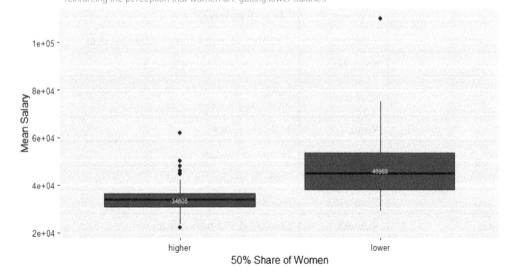

Figure 9.16 – Boxplots showing the difference for the groups
with fewer and more than 50% women enrolled

Figure 9.16 provides visual confirmation for the t-test. The group averages are different. Notice that majors with more than half of those enrolled being women have a lower mean earning, while those with more men have higher-paying salaries reported in the data. If we calculate the averages, the difference is a stunning 35.8%—a considerable gap.

With that, we have reached the end of the exploration and visualization step. We have a better understanding of the dataset, even though many other questions could be created if required by the client. Next, we will compose an analysis report to close this EDA.

Analysis report

Not all Data Science projects will be about modeling data. Most of the time, data scientists are performing data wrangling and exploration to answer business questions and extract good insights to support the decision-making process.

Sometimes, the project itself will be about the EDA, like this one we just did. In those cases, a business report can be written, highlighting the findings. Next, the report for this EDA is presented.

Dataset: *College Majors*, available at `https://tinyurl.com/4yx2dhpe`

Dimensions: 173 observations, 21 variables

Data dictionary:

Rank	Type	Rank by median earnings
Major_code	Categorical	Major code, FO1DP in ACS PUMS
Major	Categorical	Major description
Major_category	Categorical	Category of major from Carnevale et al.
Total	Numeric	Total number of people with major
Sample_size	Numeric	Sample size (unweighted) of full-time, year-round only (used for earnings)
Men	Numeric	Male graduates
Women	Numeric	Female graduates
ShareWomen	Numeric	Women as share of total
Employed	Numeric	Number employed (ESR == 1 or 2)
Full_time	Numeric	Employed 35 hours or more
Part_time	Numeric	Employed less than 35 hours
Full_time_year_round	Numeric	Employed at least 50 weeks (WKW == 1) and at least 35 hours (WKHP >= 35)

Unemployed	Numeric	Number unemployed (ESR == 3)
Unemployment_rate	Numeric	Unemployed / (Unemployed + Employed)
Median	Numeric	Median earnings of full-time, year-round workers
P25th	Numeric	25th percentile of earnings
P75th	Numeric	75th percentile of earnings
College_jobs	Numeric	Number with job requiring a college degree
Non_college_jobs	Numeric	Number with job not requiring a college degree
Low_wage_jobs	Numeric	Number in low-wage service jobs

Figure 9.17 – Description of the variables in the dataset

Missing data: There is only one entry with missing values for the `Men`, `Women`, `ShareWomen`, and `Total` variables. The major is *FOOD SCIENCE* (code: 1104).

Report

The EDA shows a complete dataset, with missing values found only for one major and representing less than 1% of the data. Therefore, it was excluded from the analysis to preserve the original data only, without adding calculated numbers.

The descriptive statistics brought good insights, demonstrating that the data has high variances due to the presence of outliers. On average, there are more women enrolled in college than men, with a split of 52% to 48%, respectively. Some majors have up to 96% of those enrolled being women, such as *Early Childhood Education* and *Communication Disorders Sciences and Services*. Regarding employment, the unemployment rate is around 6%, on average, with a deviance of 3%, presenting high variance. The annual earnings float between $30,000 and $52,000 for this dataset.

Looking at the jobs that require a college degree compared to those that do not, we see a balance if we take an overall average, even though there are some areas such as Education, Engineering, and Computers & Mathematics that are more specialized, thus having a higher percentage of jobs requiring college instruction. Considering the more specialized majors, many do not carry higher median salaries, besides those related to STEM. The Engineering category is the top-paying one, offering more than $10,000 in median earnings than the second post, Computer & Mathematics, which is also related to STEM.

When comparing women and men and how gender affects the measurements, the data shows that the majors with more women enrolled don't have a higher unemployment rate, but have more low-wage jobs related to them. On an average comparison test, the majors with more than 50% of those enrolled being women showed that the earnings are approximately 36% lower than those with fewer than 50% women.

Next steps

For the sequence of the work, if modeling is required, attention needs to be paid to the presence of many multicollinear variables, which can affect the reliability of linear regression, for instance.

The preceding text was an example of how one could write an analysis report. With this, the EDA project comes to an end and it would be ready for delivery. Likewise, *Part 2* of the book is also finished. In the next chapter, we will start *Part 3*, which is all about data visualization. In the next three chapters, we will learn a lot more about the `ggplot2` library.

Summary

In this chapter, we went over an EDA project, beginning with the load of the data to RStudio up to an analysis report.

After loading the data, we started to understand the shape of the dataset and the data types, and we did a transformation of some variables to `factor`. Moving on, we cleaned the data of missing values and started the exploration and visualization part. This began with a checkup of the descriptive statistics, then we looked at the distributions of the data and outlier detection. The sequence was to look at a bivariate chart and a pair plot that shows the correlations and scatterplots, allowing one to understand the relationship between the variables and start to get a feel of the best ones for modeling.

Next, we started to ask questions to lead our exploration, always answering them with data and statistical tests. Finally, closing the chapter, we presented an analysis report example, highlighting the findings in text form.

Exercises

1. What is the R package containing eight core libraries used in this project?

2. Why is it important to know the variable types, and which function can we use to see that?

3. How can we pull the descriptive statistics?

4. Why is the plotting of histograms and boxplots important?

5. What is the approach used in this project to lead the data exploration?

Further reading

- Pair plots using GGally: `https://tinyurl.com/34zv2feh`

- `corrplot` introduction: `https://tinyurl.com/4sybeaty`

- Dataset: `https://github.com/fivethirtyeight/data/tree/master/college-majors`

- GitHub page with code for this chapter: `https://tinyurl.com/8b7jj28f`

Part 3: Data Visualization

This part includes the following chapters:

- *Chapter 10, Introduction to ggplot2*
- *Chapter 11, Enhanced Visualizations with ggplot2*
- *Chapter 12, Other Data Visualization Options*

10

Introduction to ggplot2

In *Part 3* of this book, beginning now, the subject is data visualization. For the next three chapters, we will learn about **ggplot2**, enhanced visualizations such as interactive plots, maps, and 3D graphics, as well as how to plot graphics in Microsoft Power BI using R code.

In this chapter, more specifically, we will concentrate on learning the basics of ggplot2, a core library from the tidyverse package, to build powerful and highly customizable visualizations. The library was created by Hadley Wickham in 2015 and has evolved a lot since then, with 18 new releases. It is built on the premises of the grammar of graphics, a concept that states that a graphic must be created in layers, also called grammatical elements for this purpose.

In summary, this chapter will outline the following:

- The grammar of graphics
- The basic syntax of ggplot2
- Plot types

Technical requirements

Once again, we will use the mtcars dataset, from Motor Trends.

All the code used in this chapter can be found in the book's GitHub repository: https://github.com/PacktPublishing/Data-Wrangling-with-R/tree/main/Part3/Chapter10.

The libraries in this chapter are as follows:

```
library(tidyverse)
library(datasets)
library(patchwork)
data("mtcars")
```

The grammar of graphics

Textual communication is supported by a set of rules and elements to help us build phrases and express ideas. Grammatical elements such as nouns, verbs, adjectives, prepositions, and others are what make it possible to form sentences, which are used for the purpose of communicating a message. With that in mind, in 1999, Leland Wilkinson wrote the book *The Grammar of Graphics*, where he made an analogy between the way we write and the way we build graphics. If we think about that for a minute, it makes sense, since both structures – text and graphics – serve the purpose of communicating an idea or a message through data.

The grammar of graphics has these seven elements, which we will go over one by one: **data**, **geometries**, **aesthetics**, **statistics**, **coordinates**, **facets**, and **themes**. The first three are fundamental, as without those, there would be no graphics.

Data

The data element means the dataset being worked on that is being used to plot a graphic.

Geometry

Geometry, in the grammar of graphics, means the type of graphic to be plotted. As previously seen, there are many kinds of graphics, each one more suitable for a purpose. Histograms are good to look at data distribution, boxplots are best to discover outliers, scatterplots are interesting to understand the correlation between two variables, and line plots are more useful for time series. So, it is a matter of choosing the best geometry based on the data to be plotted and the message to be transmitted.

Aesthetics

The aesthetic elements of a graphic are what are to be seen in a plot, including what is going to be on the x axis, what is on the y axis, the color, the fill color, the shape of the geometry, the opacity, the line width, and the line type.

Statistics

The statistics element is not always applied, but it refers to graphics where there is a need to indicate the number of bins, whether data points should be counted or summed, or even the regression type for plots where it is available.

Coordinates

There are two possibilities here – cartesian and polar coordinates. Therefore, most of the time, the good old x and y Cartesian coordinates, which is the default, will be the one used.

Facets

In data science, it is not uncommon to need to plot multiple graphics for different groups within a variable. In this case, the facets are a great help, as the same figure can be used to plot all the groups, divided into separate plots. For example, if we need to see the occurrence of sun or rain on each day of the week, we can use the facets to show one plot with Yes and No bars for each weekday in the same plotting area.

Themes

Themes are the graphical element available in `ggplot2` to bring preset configurations such as background color, presence of grid lines, and fonts.

These seven elements of the grammar of graphics drive coding in the ggplot2 library. We are going to connect the dots in the next section by seeing how code is written layer by layer to build a graphic.

The basic syntax of ggplot2

Every phrase can be broken down into grammatical elements, such as subject, pronouns, and adjectives. When read together, words will make sense, forming a sentence and delivering a message. Likewise, as seen previously, there is a grammar for graphics as well, breaking the creation of a plot down into layers that can be added together to create a visual.

To understand the basics of how to write ggplot2 code, we will follow a set of questions to walk us through the process smoothly. Whenever there is a need to use the library, return to this template until the logic is absorbed and the coding becomes natural.

To create a basic plot in ggplot2, let us answer the following questions:

- What is the dataset to be used?
- What kind of graphic will be plotted?
- What goes on the *X* axis and *Y* axis?
- What is the graphic title?

These questions can be translated into the following code:

```
# What is the dataset to be used?
ggplot(data) +
# What kind of graphic to be plotted? (scatterplot)
  geom_point(
# What goes on the X and Y axes?
    mapping= aes(x= X, y= Y) ) +
```

```
# What is the title?
  ggtitle('Title for my graphic')
```

Observe that the code will always start with the `ggplot(data)` function, which will receive the dataset as its argument. That is step one. Then, we will add a new layer, and for that, we should use the addition sign, `+`. The next layer is geometry, also known as the graphic type, which can be a histogram, boxplot, points, lines, or bars. Generally, geometry will have functions such as `geom_` and the graphic type, such as `geom_point()` in the example, meaning we are plotting a scatterplot. Within that function is where the mapping of the x and y axes is done, using `aes(x, y)`. In this part, you can also add configurations such as color, filling color, size, and opacity. Finally, we add the last layer to give the graphic a title, using `ggtitle()`.

The preceding code snippet covers the three fundamental elements, according to the grammar of graphics – data, geometry, and aesthetics.

Let's see that in action using a dataset. First, we must create it, using random numbers – a distribution of 20 uniform numbers and a distribution of 20 normally distributed numbers:

```
# Create a sample dataset
df <- data.frame(
  var1 = runif(20),
  var2 = rnorm(20) )
```

Then, the scatterplot (*points*) can be built using the questions template, as previously explained in this section:

```
# What is the dataset to be used?
ggplot(df) +
# What kind of graphic?
  geom_point(
# What goes on X and Y?
    mapping= aes(x=var1, y=var2 ) ) +
# What is the graphic title?
  ggtitle('My first ggplot2 graphic')
```

Here is the resultant visualization:

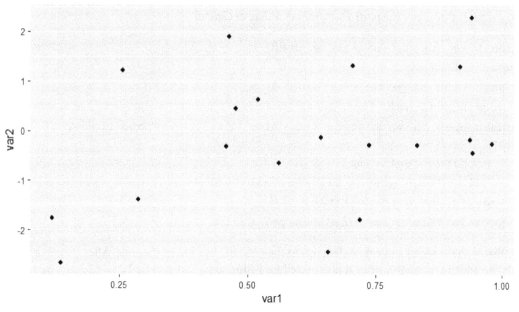

Figure 10.1 – A basic points plot created with ggplot2

We have learned the basic syntax and become intuitive about how to create a simple graphic. Now, it is time to raise the bar and learn other kinds of graphics, as well as add other elements to our plots, making them more professional looking.

Plot types

In *Chapter 3*, when studying basic data visualization, we used the *mtcars* toy dataset. In this chapter, we will go back to it, now being able to explore the capabilities of the ggplot2 library, as well as compare it to the base R plots previously created.

To load the dataset into an RStudio session and code along with this book, use the `data("mtcars")` code. To make the code more generic and transferable to other data frames, I will call the dataset `df`.

Histograms

Histograms are created using the `geom_histogram()` function. As usual, the same questions are applicable here to write the code:

- What is the dataset? *df.*

- What is the kind of graphic? *Histogram.*

- What goes on *x*, and what is the color, fill color, and number of bins? *Miles per gallon, with 20 bins.*

- What is the title of the plot? *Histogram of Miles per Gallon*:

```
# What is the dataset to be used?
ggplot(df) +
# What kind of graphic?
  geom_histogram(
# What goes on x, what is the color, fill color and number of
bins?
    mapping= aes(x= mpg), bins= 20,
    color="lightgray", fill="royalblue") +
# What is the graphic title?
  ggtitle("Histogram of Miles per Gallon")
```

Note that we have added some new elements to the aesthetic portion of the code. After mapping the *x* axis, we added `bins=20`, since it is expected for histograms, and also `color="lightgray"` and `fill="royalblue"` to set up the border and filling color of the bars. The code will display the graphic shown in *Figure 10.2*.

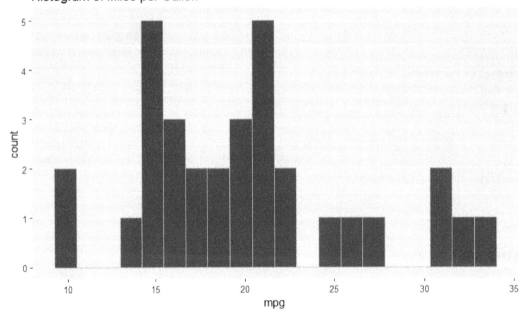

Figure 10.2 – Histogram of miles per gallon

The graphic shows that we have some bins without observations, which could be interpreted as good places to divide the data into groups of MPG, but, if that does not make sense to our analysis, the gaps can be filled by decreasing the number of bins.

Boxplot

The boxplot is a good ally to find outliers, as well as it is a sort of visual T-test, being an excellent resource to compare groups' averages. To create it, use the `geom_boxplot()` geometry function.

From now on, I won't keep repeating the questions template, just for the sake of space, but keep them in mind when writing ggplot2 code. The snippet to follow is for a boxplot of MPG. Since it is a univariate plot, you only need to provide one axis, which will be *y*, so the box is positioned vertically:

```
# Boxplot of MPG
# Dataset
ggplot( df ) +
# Geometry, Y and filling color
  geom_boxplot( aes(y=mpg), fill="royalblue" ) +
# title
  ggtitle("Boxplot of Miles per gallon")
```

As a result, the following graphic is displayed:

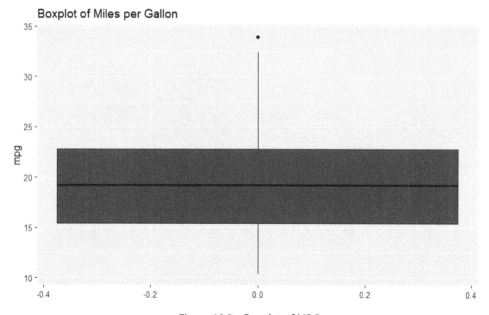

Figure 10.3 – Boxplot of MPG

Figure 10.3 shows only one outlier for the MPG variable, somewhere around 34 MPG.

Boxplots are also a good choice if you are interested in comparing groups' averages. The result will be very similar to what you get when using a statistical T-test (HAIR Jr. et al, 2019), but you will have a visual return with this graphic. Let's compare the effect of having or not having a V-shaped engine on the MPG variable. Note that this time we are providing the `aes()` aesthetics element with the *x* and *y* axes, where `x= factor(vs)` indicates that the variable for the engine shape (`vs`) should be read as categories by ggplot2, not as numbers. The `labs()` function, now introduced, is used here to rename the *x* label `Engine shape`:

```
# Dataset
ggplot( df ) +
# Geometry, X, Y and filling color
  geom_boxplot( aes(x= factor(vs), y=mpg), fill="royalblue" ) +
# overwrite the X label
  labs(x="Engine shape") +
# Title
  ggtitle("A comparison between V-shaped vs line-shaped engines
and the effect on MPG")
```

The comparison coded previously outputs *Figure 10.4*.

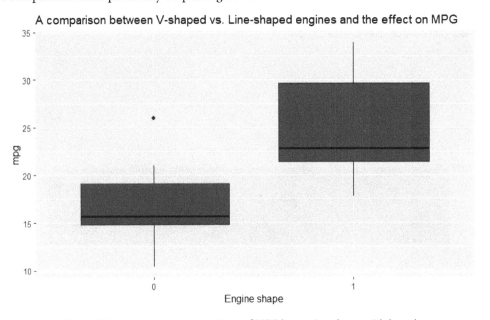

Figure 10.4 – An average comparison of MPG by engine shape with boxplots

According to the test, the cars with line-shaped engines (1) are more economic than the V-shaped equipped cars.

Scatterplot

The scatterplot is also known as a points plot. Ergo, that was the name of the geometry chosen by the library's creators. Use geom_point () to create a scatterplot.

This graphic type is very useful for understanding relationships between variables and correlations. Observe the code that follows. There is the basic ggplot (df) to create the figure, the geometry will be geom_point (), the aesthetics provided to the geometry are the x and y axes and the color, size (15 means squares), and shape values of the markers, and we added the alpha argument this time, which relates to the opacity of the points. Then, we used the labs () function to rename x and y, as well as to include title and subtitle in the graphic:

```
# Scatterplot weight versus mpg
ggplot(df) +
  geom_point ( aes(x= wt, y= mpg),
               color= "royalblue", size=4, shape=15, alpha=0.7 )
+
  labs(x= "Weight of the cars", y= "Miles per gallon",
       title= "How the weight affects MPG in cars?",
       subtitle= "As the weight increases, the car will make
less MPG")
```

The plotted graphic is subsequently displayed:

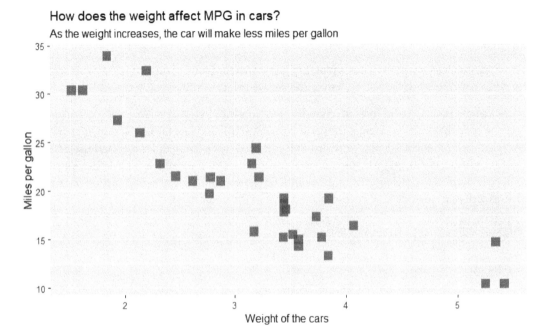

Figure 10.5 – A customized scatterplot of MPG versus weight

The plot from *Figure 10.5* looks very professional. Observe that the customizations made increase the readability of the graphic. The addition of subtitles helps us to know what to expect from the data, which is a negative correlation between the variables. Opacity makes it easier to see where the points overlap. This is especially good when using datasets with more observations, where there are a lot of overlaps between points.

Bar plots

There are many kinds of graphics, but not many are as simple to read as a bar graphic. For categorical plots, bar or column plots are essential, showing counts or values for each category represented. To create such graphics, use the geom_bar() or geom_col() geometry functions. There is a slight difference between them, which will be explained next.

When using the geom_bar() function, we are usually looking at a single variable. If we want to plot a count of observations by transmission type (am), we can call the geometry function and pass factor(am) as the categories to be counted and rename the label x using labs(). Note one specificity – this time, we passed fill=factor(am) inside the aesthetics function. That small change makes a significant difference because it will make ggplot2 understand that we want to fill one color for each category. So, the `fill` argument within `aes()` means one color per category, while the `fill` argument outside `aes()` means the same color for everyone. Let's see that in action:

```
# Bar plot
ggplot(df) +
  geom_bar( aes(x= factor(am), fill= factor(am) ) ) +
  labs(x="Automatic(0) | Manual(1)") +
  ggtitle("Count of observations by transmission type")
```

Figure 10.6 is what comes out of the preceding code.

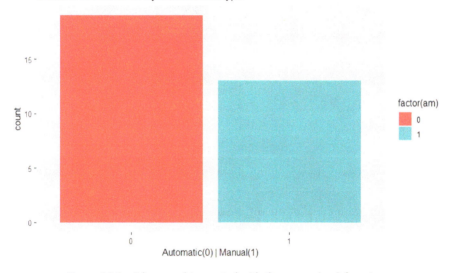

Figure 10.6 – A bar graphic created with the geom_bar() function

In this dataset, there are more cars with automatic transmission, which is a bit surprising for a 1974 sample of cars.

The bar plot geometry comes with the count statistic associated with it. Hence, if we need to change that to mean, for example, we will have to use the statistics function, which is one of the seven grammatical elements. It is basically the same plot, except we are adding the `stat= "summary"` arguments and indicating that we want to use the mean calculation (`fun= "mean"`):

```
ggplot(df) +
# statistic calculation - mean value
  geom_bar(aes(x= factor(am), y= mpg, fill= factor(am)),
           stat = "summary", fun="mean")
```

The result will be a bar graphic, with the bar being the average MPG value for each transmission type.

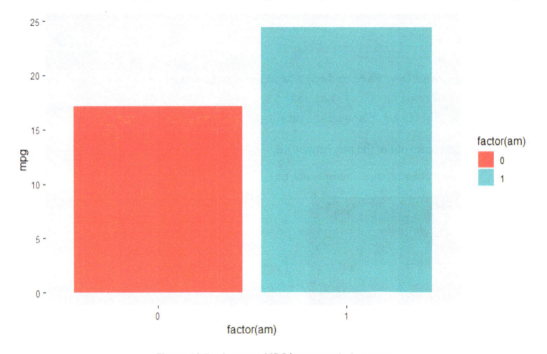

Figure 10.7 – Average MPG by transmission type

Remember that we did not include a title or rename the *x* axis; thus, *Figure 10.7* won't display those updates.

There is yet another syntax, where instead of calling geom_bar(), we will call the stat_summary() function, providing it with the aesthetics, plus the function to be used for the *y* value (fun= "mean"), and the geometry required (geom= "bar"):

```
# Bar plot with two variables
# Another syntax, using stat
ggplot(df) +
# statistic calculation - mean value
  stat_summary(aes(x= factor(am), y= mpg, fill= factor(am)),
fun="mean", geom="bar")
```

The resultant plot will be equal to *Figure 10.7*.

The other way to create a bar or column plot is by using geom_col(). This function requires x and y axes and returns bars with the summed amount of the y-axis variables by category:

```
# Column plot for MPG by transmission type
ggplot(df) +
  geom_col( aes(x= factor(am), y=mpg, fill=factor(am) ) ) +
  labs(x="Automatic(0) | Manual(1)")
```

The only change in the code compared to the bar plot is the geometry function, which now is geom_col(). The code will result in the following graphic. Be aware that values on the y axis are summed, and that amount does not make much sense for our analysis. What we would prefer in this case would be a count or another statistic, such as a mean or median value.

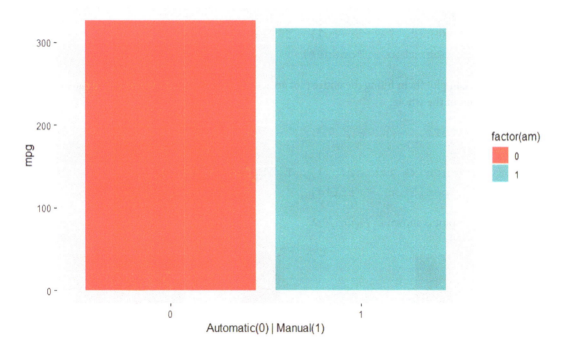

Figure 10.8 – A column plot of MPG by transmission type

A workaround for this is to create a subset with the mean values of the groups and then plot it using geom_col(), as we did many times in the EDA of *Chapter 9*.

For the bar plots, we can also change the position argument to make it stacked, filling the entire axis or side by side. So, next, while checking what is the dominant type of engine by cylinder in the dataset, we will plot three bar plots and compare their styles. The code to construct this is similar to past code, but here, we will add the `position="stack"` argument in the `geom_bar()` function:

```
# Bar plot stacked
ggplot(df) +
  geom_bar( aes(x= factor(cyl), fill=factor(vs) ),
            position = "stack")
```

Similarly, we can plot the bars side by side:

```
# Bar plot side
ggplot(df) +
  geom_bar( aes(x= factor(cyl), fill=factor(vs) ),
            position = "dodge")
```

Alternatively, we can plot them filling the entire plot area, as a 100% bar, and the respective categories' representation out of the whole:

```
# Bar plot fill
ggplot(df) +
  geom_bar( aes(x= factor(cyl), fill=factor(vs) ),
            position = "fill")
```

The result for each plot is shown in *Figure 10.9*.

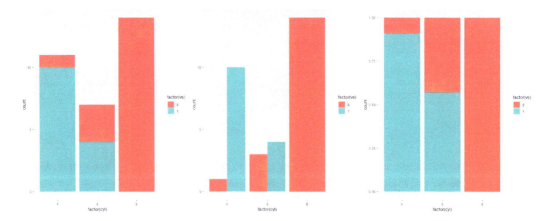

Figure 10.9 – The three position options for a bar plot

There are three types of positions for bar plots, with the cylinders on the *x* axis and the observation counts on the *y* axis. The coral color is for V-shaped engines and teal is for straight-line engines. Observe that the engines in line dominate the four-cylinder models on the left-hand-side bars, the six-cylinder engines are almost equally split, and the eight-cylinders engines are all V-shaped in this dataset. It is hard to tell which option is the best, as this may vary from one project to another. So, you can choose the one that fits better with your project. In this case, with just two categories, the first plot on the left is more concise and delivers a clear message.

Next, let's learn about line plots.

Line plots

Line plots are observed to show progression over time. Let's put the *mtcars* dataset aside for now and suppose that there is a dataset with monthly car sales, and we want to visualize that information – the line plot would be one of the best indications.

It is easy to create a dataset with random numbers for months and car sales to exemplify our case. See the dataset created in the table of *Figure 10.10*.

month	sales	sales2
1	13,709	15,277
2	5,646	3,066
3	3,631	1,466
4	6,328	6,995
5	4,042	3,126
6	1,061	29,221
7	15,115	26,845
8	946	14,521
9	20,184	3,373
10	627	19,594
11	13,048	1,891
12	22,866	13,361

Figure 10.10 – A dataset created with random numbers

To create a line plot, use the `geom_line()` geometry function, add x and y to the aesthetics, and also an argument called `group=1`, which is needed to group the sales numbers as a single group so that the library knows which points to connect. These are complemented by `size`, `color`, and title.

```
# Simple Line plot
ggplot(sales) +
  geom_line( aes(x=month, y= sales, group=1),
             size=1, color="darkgreen") +
  ggtitle("Car sales throughout the months")
```

The code displays the following graphic:

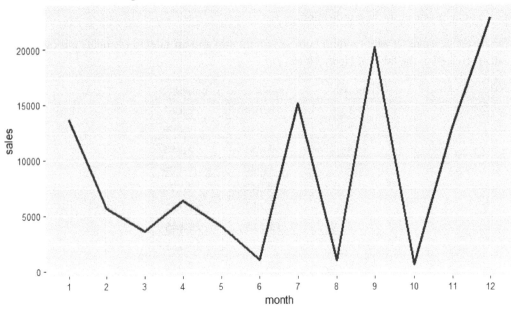

Figure 10.11 – A simple line plot

There will be times when we need to plot multiple lines in the same figure for comparison. Imagine that there are sales numbers for 2 years in our dataset. In this case, we can use similar code, just by adding another `geom_line()` for the new variable to be added:

```
# Line plot with two variables
ggplot(sales) +
  geom_line( aes(x=month, y= sales, group=1, color="sales year
  1"),
```

```
                size=1, linetype=2) +
  geom_line( aes(x=month, y= sales2, group=1, color="sales year
2"),
                size=1, linetype=1) +
  ggtitle("Car sales throughout the months - Two year
comparison")
```

Note that the `color` argument was placed within the `aes()` function, and we provided a name for the legend. The code will show the following graphic in *Figure 10.12*.

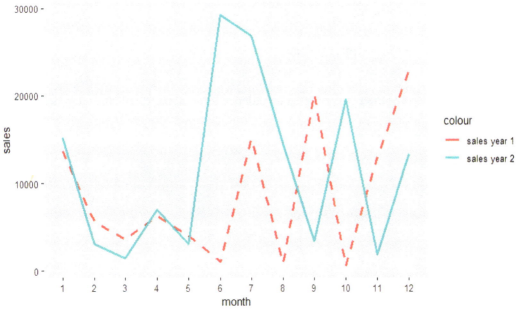

Figure 10.12 – A line plot of two series

The graphics from ggplot2 are very elegant, and there are more cool features to be presented. Let's move on.

Smooth geometry

The geom_smooth() function, as per the documentation, calculates a smoothed line that helps us to see trends in the points, using methods such as linear regression, a general linear model, and polynomial regression to create a trend line that helps us interpret the graphic.

The use is similar to other geometries. In this case, we will plot a smooth line on top of a scatterplot of the effect of horsepower over MPG. Note that we add geom_point() first and then geom_smooth(), inputting the same *x* and *y* axes to the function. By default, the method used is a polynomial function ("loess"), but there is linear regression ("lm"), generalized additive models ("gam"), robust linear models ("rlm"), and generalized linear models ("glm"):

```
# Smooth line geometry
ggplot(df) +
  geom_point( aes(x= hp, y= mpg, color= factor(vs)) ) +
  geom_smooth( aes(x= hp, y= mpg), method='loess')
```

The result shows the smoothed curve with a confidence interval band. The function is a very impressive tool.

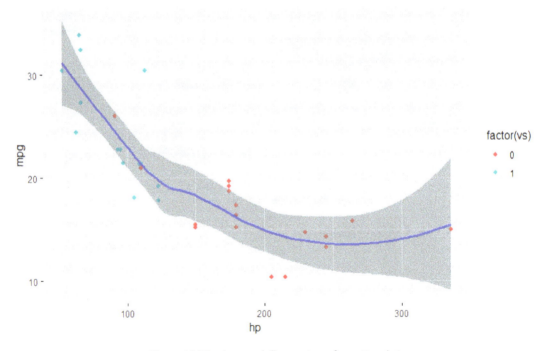

Figure 10.13 – A smooth line on top of a scatterplot

The smoothed line on the plot from *Figure 10.13* helps us to easier see the effect that horsepower has over MPG in this data. Up to 200 hp, there is a clear downward trend in the MPG, but after that, the effect starts to fade, and the cars won't lose much more efficiency as they increase the engine power.

Hadley Wickham, the creator of ggplot2, says that we should notice that the previous code is not as efficient as it could be. The axes are being repeated twice, since we added two geometries in the plot. That can be easily fixed if we pass x and y axes in the aesthetics function, together with the ggplot() function that holds the dataset. This simple change will make the library apply those axes to all the geometries added. Let's re-code the last snippet and see how it looks:

```
# Smooth line
ggplot(df, aes(x= hp, y= mpg) ) +
  geom_point(aes(color= factor(vs))) +
  geom_smooth(method='loess')
```

The result is a graphic identical to *Figure 10.13*.

Themes

Another grammatical element to support graphic creation in ggplot2 is the theme. Themes are preset visual configurations that you can add to code as a layer, creating a plot that makes more sense with the style of the project or (why not?) the one you like the best.

To add a theme, you can choose one of the following functions and add that as a coding layer: theme_bw(), theme_classic(), theme_light(), theme_dark(), theme_gray(), theme_linedraw(), theme_minimal(), and theme_void():

```
# Adding Theme BW
g = ggplot(df) +
  geom_bar( aes(x= factor(am) ), fill= 'royalblue' )
```

The following figure shows the themes and respective plots:

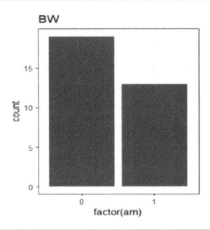

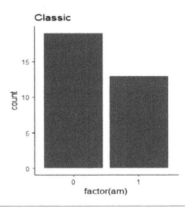

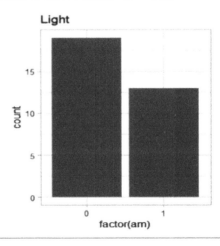

`g + theme_dark()`

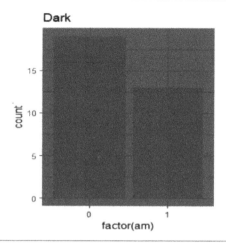

`g + theme_gray()`

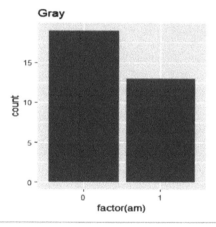

`g + theme_linedraw()`

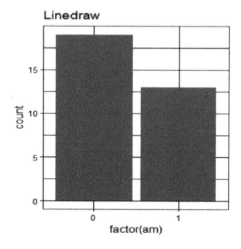

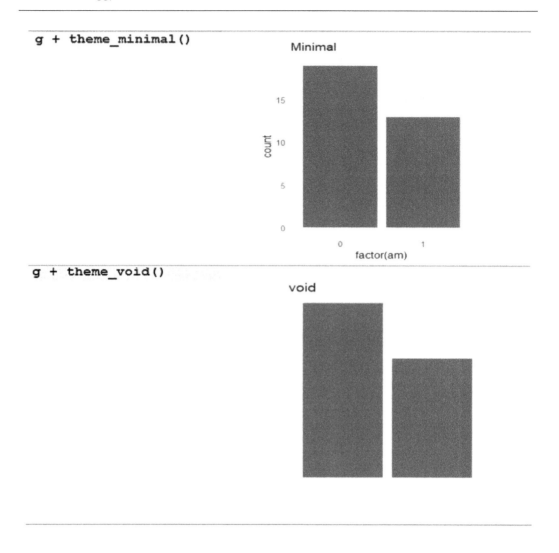

Figure 10.14 – Themes available in ggplot2

Figure 10.14 shows how each theme looks.

There is a lot more that ggplot2 offers. It is a complete library for plotting of any kind, and if I tried to show everything, it could fill an entire book. Therefore, I highly encourage you to examine the *Further reading* section and look at the documentation to learn all the amazing things that this package can do.

In the next chapter, we will delve a little deeper into visualization to create some enhanced types of graphics, such as facet grids, maps, and 3D graphics.

Summary

In this chapter, we studied one of the main graphic packages in marketing. The ggplot2 library is capable of so much that it was even translated into other languages, such as Python.

We began the chapter discussing the interesting theory of the grammar of graphics, using an analogy of textual grammatical elements and looking at the elements needed to construct and plot a good visualization. ggplot2 was built on top of that concept, enabling analysts to code a graphic by layers, adding one piece at a time. We then introduced a template of questions to help organize our thinking when creating code: (1) start with a dataset, (2) choose a geometry, (3) provide axes and aesthetics, and (4) add a title, labels, statistics, and themes.

After familiarizing ourselves with the syntax, we studied the code for the most commonly used types of graphics, such as histograms, boxplots, scatterplots, bar plots, and line plots. Then, we introduced smooth geometry, which helps us to create smoothed lines and even linear regression lines on top of graphics. Finally, we ended the chapter by showing you how to add a theme to a graphic.

In the next chapter, we will look at data visualization.

Exercises

1. What are the three fundamental elements of the grammar of graphics?
2. What is the main plotting function of ggplot2 that holds the dataset?
3. What are geometries and why they are important to a graphic?
4. List five geometry functions.
5. What is the function used for aesthetics? Does it go within another function?
6. What is the difference between placing a color argument inside or outside an aesthetics function?
7. How can we add a theme to a graphic?

Further reading

- Documentation and cheat sheet for ggplot2: `https://ggplot2.tidyverse.org/`

- A high-level view of the grammar of graphics: `https://www.youtube.com/watch?v=ieDyz7xUK4k`

  `https://tinyurl.com/mrxnz75p`

- The grammar of graphics – University of Chicago: `https://cfss.uchicago.edu/notes/grammar-of-graphics/`

- Line types and marker types for ggplot2: `https://ggplot2.tidyverse.org/articles/ggplot2-specs.html`

- Geom_smooth documentation: `https://ggplot2.tidyverse.org/reference/geom_smooth.html`

- R code in GitHub for this chapter: `https://tinyurl.com/3tkj9v5k`

- Adding labels to a bar graph: `https://r-graphics.org/recipe-bar-graph-labels`

11

Enhanced Visualizations with ggplot2

Data visualization is an art. Choosing the right graphic type, the right x and y variables, and the right colors, shapes, and titles can be challenging. We must be careful not to make our graphic too crowded with information or too lacking in information.

Sometimes, it is necessary to add other resources to the plot that will help us to deliver the right message, or at least to make it easier for the audience to understand. This is when additions such as facet grids, interactivity, and maps can be helpful.

In this chapter, we will go over some of these additional elements that can improve the readability or the interpretivity of a graphic. We will start with the facet grids, one of the grammatical elements that we still haven't covered; then, we move forward to study 3D plots and when to use them. After that, we will learn about map plots, a valuable tool to have these days, as the world is more connected every day. And finally, there will be a section about how to add interactivity to plots.

We will cover the following main topics:

- Facet grids
- Map plots
- Time series plots
- 3D plots
- Adding interactivity to graphics

Technical requirements

We will use the *diamonds* dataset, which is including in the ggplot2 library.

All the code can be found in the book's GitHub repository: `https://github.com/PacktPublishing/Data-Wrangling-with-R/tree/main/Part3/Chapter11`.

The libraries in this chapter are as follows:

```
library(tidyverse)
library(lubridate)
library(datasets)
library(patchwork)
library(plotly)
data("diamonds")
```

Off we go.

Facet grids

Facet grids create a figure in the form of a matrix of rows and columns to plot multiple graphics side by side. Those graphics are subplots of one or more variables, facilitating the visualization of the relationship of a variable with others separately. In summary, facet grids show small plots representing a subgroup of the data.

We can see what a facet grid looks like using the *diamonds* dataset, which is built into **ggplot2** (type `?diamonds` into R's console for the documentation). This data has the cuts, dimensions, colors, prices, and other attributes of 54,000 diamonds. If we want to see a scatterplot of the prices by carat, the graphic will look busy, as we can see in *Figure 11.1*. Notice that it is difficult to see the trends and relationships for each cut type, such as Fair or Good. They will be hidden under other points. What we see is the general trend and relationship for the entire dataset.

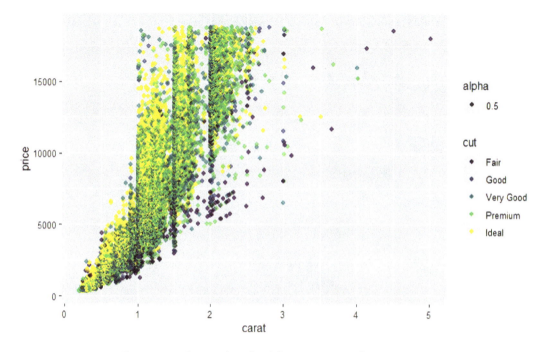

Figure 11.1 – Scatterplot of weight versus price of the diamonds

Now, when a facet grid is implemented, the visualization becomes easier because it is possible to see relationships separated by group. As usual, the syntax is the same. We start with the main ggplot(df) function, then we add the point geometry, with the aesthetics for the *x* and *y* axes. The color value is different for each cut, and the opacity is set to 50% using the alpha argument. The new addition in this case is the facet_grid() function, which can receive the rows argument (or cols, whichever you prefer) with the variables that will be used for the separation within vars():

```
# Facet grid by cut of scatterplot price by carat
ggplot(df) +
  geom_point( aes(x= carat, y= price, color= cut, alpha=0.5)) +
  facet_grid( rows = vars(cut) )
```

The result is this beautiful visualization in *Figure 11.2*, separating one subplot of cut per row.

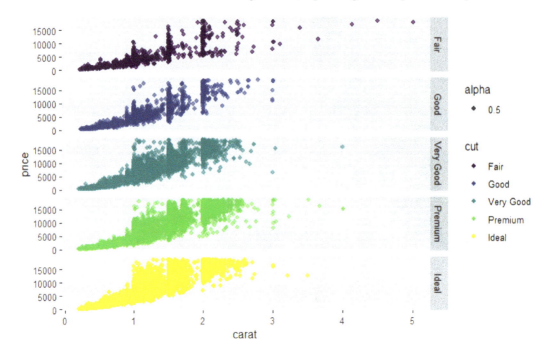

Figure 11.2 – Scatterplots in a facet grid separated cut type

You can see that the plot is much less busy now. It is easy to look at the relationship between price and weight by cut type now. Observe that the three best cuts have a high density of points in the high prices, which makes sense for this dataset. A diamond is expected to be more expensive as its weight increases.

It takes just a small change to add more plots to a facet grid. Suppose we are interested in seeing how the *clarity* of the diamond affects the price as well. We can use the same code snippet that we used previously but add the cols argument, for columns, using vars(clarity) as input:

```
# Facet grid by cut and clarity of scatterplot price by carat
ggplot(df) +
  geom_point( aes(x= carat, y= price, color= cut, alpha=0.5)) +
  facet_grid( rows = vars(cut), cols= vars(clarity) )
```

That will result in a matrix of subplots where the rows are different types of cut and the columns are different types of clarity. ggplot2 will create one subplot for each combination.

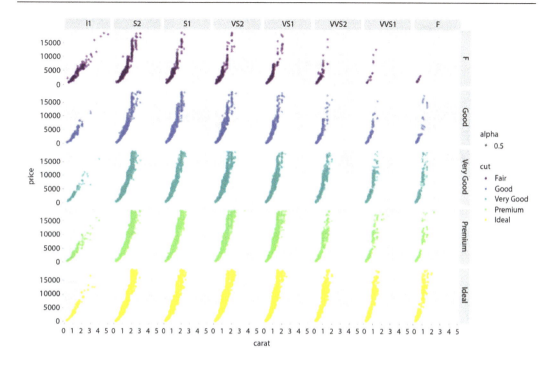

Figure 11.3 – A facet grid with rows and columns

Figure 11.3 displays each combination of the variables on a separate subplot.

A variation of the facet grid is the `facet_wrap()` function, which forces the subplots into a rectangular matrix. If we condense the `ggplot()` function and the geometry and aesthetics function to the variable g, then we just need to add `facet_wrap()` and pass the variables to create the grid within `vars()`:

```r
```{r}
# Facet wrap by cut of scatterplot price by carat
g + facet_wrap( vars(cut) )
```
```

This is what the code returns:

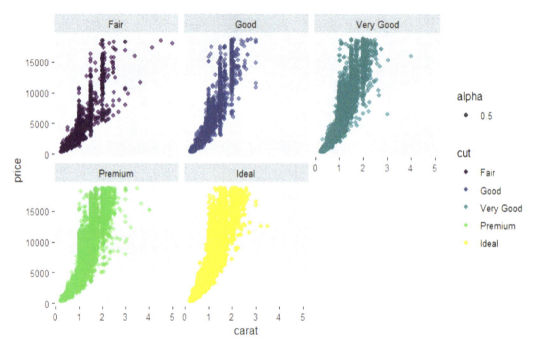

Figure 11.4 – Facet wrap by cut

Figure 11.4 shows the subplots organized in a rectangular 2x3 matrix.

To recap, all it takes to create a facet grid is to add the `facet_grid()` function to the base and geometry, providing it with the variables that will go in the rows and/ or columns. Those variables must be within `vars()`.

The next type of enhanced plot is the map plot. Let's continue.

Map plots

We live in the information era. Enormous amounts of data are created each day, from all parts of the world. Part of that data has location information attached to it (latitude and longitude), enabling the data scientists that have access to it to create visualizations using maps. Anaysis of store sales by city, state taxes collection, tourism destinations, and internet access by country are only a few examples of a large spectrum of possibilities. That is enough reason to learn how to use ggplot2 to create plots using maps.

A side note before we jump into the action is that map plots are a vast domain as well, being part of the spatial data analysis domain, which is out of the scope of this book. Here, the intention is to show the capabilities of ggplot2. To learn in more depth about map plotting, there is some material available in the *Further reading* section.

To plot a map, the geometry used is geom_map(). But before we can plot anything, ggplot2 requires us to load the map object, which will serve as the background for the plot. It will bring up a shape file to create the map drawing and then plot things over it. In this exercise, we will plot things over a map of the USA, so let's load the map of the USA, showing state borders. Let's load a data frame with latitude, longitude, and state names:

```
# Loading the map of the USA states from ggplot2
us <- map_data("state")
```

We can plot this using the geom_map() function now, using x=long and y=lat. map_id= region draws each state name associated with a latitude and longitude. theme_void() suppresses any borders or axes or labels:

```
# Plot only the USA map
ggplot(us) +
  geom_map( aes(long, lat, map_id= region), map=us,
            color="black", fill="lightgray") +
  theme_void()
```

The result is a map of the USA, as shown in *Figure 11.5*.

Figure 11.5 – USA map divided into states, plotted with ggplot2

Now that we have this, we can load another dataset, a CSV file with the states, the state capital names, latitude, longitude, and the **Gross Domestic Product (GDP)** of each state:

```
# Load the CSV file
states <- read_csv("USA_states.csv")
```

The next step is to transform the state names to lowercase because they have to match the name in the *region* column in the dataframe from ggplot2, and the state names there are all lowercase (look at *Figure 11.6*).

We learned how to do this transformation in *Chapter 4* using `mutate()` in association with the `str_to_lower()` function from the `readr` library:

```
# Names of the states to lower case
states <- states %>% mutate(state= str_to_lower(state))
```

Next, there is some code to plot the *states* dataframe, adding `geom_map()` with `longitude`, `latitude`, and `map_id= state`, since that is the variable from our *states* dataset that matches the *region* names from the ggplot2 map shape, the data frame that we called us, as seen in *Figure 11.6*. `xlim()` and `ylim()` limit the size of the plot and centralize the map of the USA on the screen. All of that is stored in the variable called us_map:

```
# Mapping the location of the state capitals
us_map <- ggplot(states) +
  geom_map( aes(longitude, latitude, map_id= state), map=us,
            color="black", fill="lightgray") +
  xlim(-130,-60) +
  ylim(20,55) +
  theme_void()
```

data frame: us

	long	lat	group	order	region	subregion
1	-87.46201	30.38968	1	1	alabama	NA
2	-87.48493	30.37249	1	2	alabama	NA

data frame: states

	state	state_cd	capital	latitude	longitude	GDP
1	alabama	AL	Montgomery	32.37772	-86.30057	247092.5

Figure 11.6 – Matching state names for the variables in the us (ggplot2) and states (csv) dataframes

What we just did was create the map again, matching it with the ggplot2 map shape. If you plot `us_map`, the result is the same as in *Figure 11.5*.

It is easy to add another layer to show where the capital cities are located. This is done simply by adding `geom_point()`. I chose `shape=24`, which plots triangles instead of circles, and the size and fill are set by the GDP variable. Therefore, the higher the state's GDP, the larger the triangle on the map:

```
# Adding the capital of the states locations.
us_map +
  geom_point( aes(x=longitude, y=latitude, size= GDP, fill=
GDP),
              shape=24 )
```

The result will look like this:

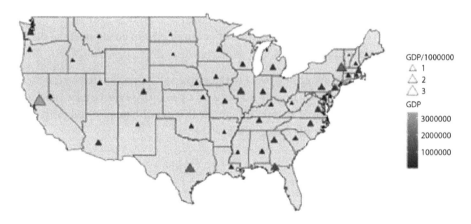

Figure 11.7 – USA map with the capital cities marked

The map in *Figure 11.7* shows the GDPs of all the states. The triangle is the location of the capital city and the bigger the triangle, the higher the state GDP. The amounts are in millions of USD, meaning that the 3,000,000 shown on the right of the plot should have six more zeroes added, thus it is three trillion dollars (`https://tinyurl.com/bdvn4dzw`).

Map plots are very useful and they cause a good impact, as it is possible to relate data to a geographical location. Next, we will go to another interesting dimension: time. We will learn about ggplot2 visualizations for time series.

Time series plots

A time series is a sequence of data points ordered by time. In a time series, the data points are measurements of any given variable throughout time, such as days, hours, months, or any other time frame. We can visualize time series using ggplot2 as long as the dataset contains a datetime variable. The best way to visualize data organized by time is with line plots. Let's set a seed so you can reproduce the same results as mine for the random numbers. Create a sample dataframe and then see how to visualize a time series:

```
# Set seed to reproduce the same random numbers
set.seed(10)
# Creating a Dataset
ts <- data.frame(
  date = seq(ymd('2022-01-01'),ymd('2022-06-30'),by='days'),
  measure = as.integer(runif(181, min=600,  max= 1000) +
sort(rexp(181,0.001))))
```

The preceding code is a data.frame object where we are assigning a sequence of dates to the name date from January 1 to June 3, 2022, with one observation by day. To create such variable, we are using the seq() function. The measure variable is composed of a uniform distribution between 600 and 1,000. Then, for each of those we have added another number from an exponential distribution to create an uptrend. The dataset is stored as ts.

Moving on, to plot the ts data using ggplot2, we use the ggplot() function fed by the ts dataset. We add line geometry with geom_line() to set x=date and y=measure, and aesthetic elements such as line size and the classic theme. We can use this as our base plot and assign the name basic_plot, which will make our life easier in the next steps:

```
# Basic plot
basic_plot <- ggplot(ts) +
  geom_line( aes(x=date, y=measure), size=0.7)+
  theme_classic()
```

The graphic plotted is shown in *Figure 11.8*.

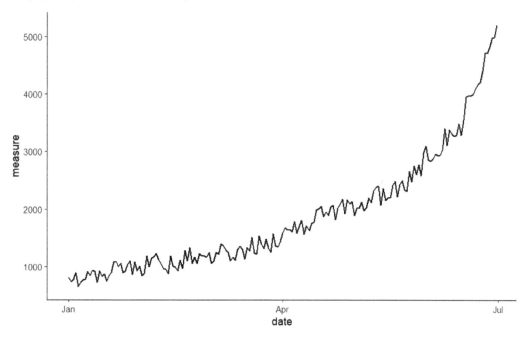

Figure 11.8 – Basic time series plot

Notice that our time series has a strong uptrend. Each month, the measure gets higher and higher. But since only three months are displayed on the *x* axis, it is hard to tell at a glance what year is represented by the data in the graph. As we have already seen when we discussed the ggplot2 library, it is a matter of adding layers to make changes to it. Let's change the *x* axis display to show the complete date just by adding the `scale_x_date()` function and changing the `date_labels` argument. You can find a complete list in the link in *Further reading,* but here are the most common string patterns for datetimes in R: `%d` for day, `%m` for 2-digit month numbers, `%b` for abbreviated month names, `%B` for full month names, `%W` for week numbers, `%y` for 2-digit years, and `%Y` for 4-digit year numbers:

```
# Changing the X axis label format to 4 digit year (%Y),
abbreviated month (%b) and day (%d)
basic_plot + scale_x_date(date_labels = "%Y %b %d")
```

Figure 11.9 displays the result.

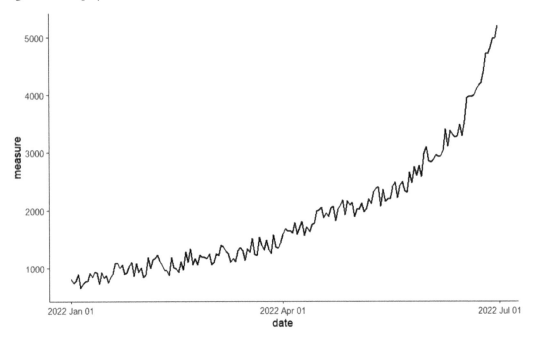

Figure 11.9 – Time series plot with x axis labels customized

Now we can see the complete date, but we still can't tell for sure what the date is when the uptrend gets steeper at the end of the series. Let's add more information to the *x* axis. Using the same `scale_x_date()` function as before, the `date_breaks` argument allows us to set up the interval of the *x* axis ticks. In this case, we are using a 2-week interval, but it could be 1 day, 1 month, or any other period:

```
# Breaks by bi-weekly month
basic_plot +
  scale_x_date(date_breaks = "2 weeks", date_labels = "%W %y")
```

The code will produce the next graphic.

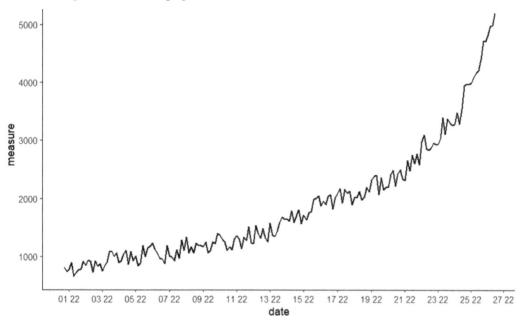

Figure 11.10 – Time series plot with customized x axis ticks

As we can see in *Figure 11.10*, it is clearer now that the curve becomes steeper after the 25th week of 2022. It is also a simple task to change the date_labels argument value to %m %d and get the actual date, which is the week beginning June 20. Then, we can have a closer look at that part of the graphic using the following code:

```
# Closeup on date after 06/01
basic_plot +
    scale_x_date( limit=as.Date( c("2022-06-01", "2022-07-01") ),
                  date_breaks = "1 week", date_labels = "%m %d")
```

This code uses scale_x_date once again, and the limit argument will determine the period we want to focus on. As in the previous analysis, we want to focus on dates around June 20. Thus, we will plot from June 1 to July 1. The other argument is date_breaks, set to 1 week, and the labels used were %m for the month digits and %d for the day digits. The code will give us the graph shown in *Figure 11.11*.

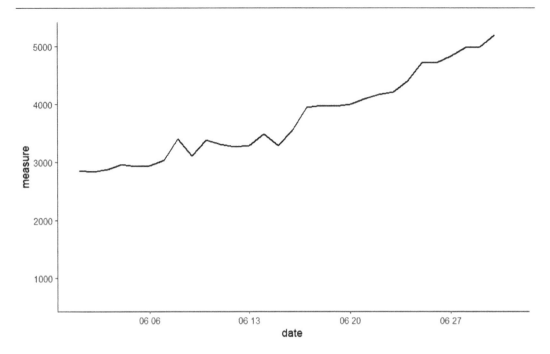

Figure 11.11 – Focus on specific dates in a time series

The graphic from *Figure 11.11* shows that the measure was growing at a slow rate until around June 15, when it became much faster. In a real-life project, we would combine this insight with more data to determine what happened to drive such an increase, but I am sure you understand how powerful these functions for plotting time series can be.

With the content from this section, it is possible to plot, customize, and slice time series graphics. The next section will cover 3-dimensional plots, which library to use to create them, and when to use them.

3D plots

3-dimensional plots are beautiful. Very often, they create a good impression with their audience, but the truth is that they are not the best type of graphic to use. To plot a 3D graphic on a 2D space, such as on a computer screen or on paper, the third dimension will have to simulate depth that does not exist. It is not recommended that you plot in 3D very often, as in general, a good old 2D plot will be the simplest and best option.

Sometimes, though, looking at 3D plots can be useful. Cases such as surface graphics, which can represent the surface of a given place, such as a mountain, can be interesting.

3D graphics can be created using the **plotly** library in R (loaded with `library(plotly)`). Let's create a random surface and plot it. The surface graphics require the input data to be a matrix, thus we are creating one and then using the `plot_ly()` function, passing the `z= ~surface` argument to it to indicate that we want a 3D graphic. Remember that x and y are for 2D, and z is the third dimension. To finish, we add a surface to the basic function.

Next, we create an 80x20 matrix of integers in a uniform distribution (`runif()`):

```
# Creating a random surface
surface <- matrix(as.integer( sort( abs(runif(160,90, 180)) )
),
          nrow=80, ncol=20)
```

Then we plot the 3D surface with the next code:

```
# Plotting the surface
plot_ly(z=~surface) %>% add_surface()
```

As the result, it will show the following graphic.

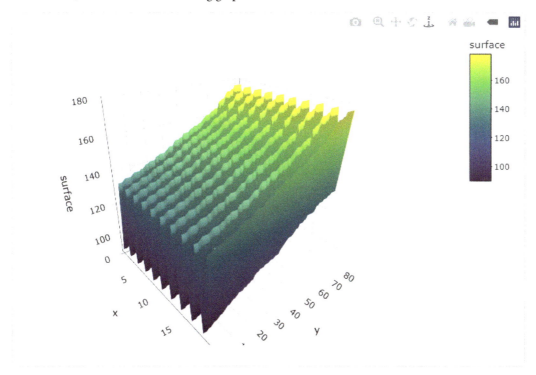

Figure 11.12 – 3D surface plot

The simulation of terrain with 3D plots can be very useful for many industries. Look how we get a good idea of the surface in *Figure 11.12*.

Another good use of 3D plots is when two variables have that are points very close to each other, making it hard to see the differences or to split them into groups. In this case, the separation may be happening in another dimension, or in other words, in another variable. Let's simulate another scenario where we have `var1` and `var2` with a minimal difference between them. The code that follows has two random normal distributions where `var2` has a value of `var1 + 0.1`:

```
# Variables
var1= rnorm(20, mean=25, sd=5)
var2= var1+0.1
```

Next, here is a `data.frame` object created from those variables, with a third one (`var3`), where the separation happens, plus an identifier (`var4`) of the two groups, repeating the ID of each group 10 times:

```
#Data frame
my_data <- data.frame(var1= var1,
                      var2= var2,
                      var3= 1:20,
                      var4= rep(c('A','B'),each=10) )
```

Here is the code to plot the 2D graphic, a scatterplot created using `ggplot2`:

```
# Plot 2D
ggplot(my_data) + geom_point( aes(x=var1, y=var2, color=var4) )
```

The code yields the next graphic.

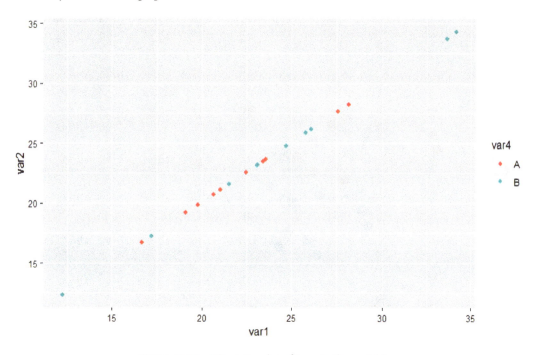

Figure 11.13 – 2D scatterplot of two similar variables

Notice that it is difficult to see the separation of A and B in *Figure 11.13*, given that they are very similar, indeed, almost equal. But let's make a 3D plot now. To produce a 3D plot, the function is `plot_ly()`, provided with the `my_data` data frame, the x, y, and z variables, the identifier variable `var4`, and the colors for each group. We complete it with the `add_markers()` function, to add the points to the plot. Observe that the code for plotly requires the pipe signal, not addition.

```
# Plot 3D
plot_ly(my_data, x=~var1, y=~var2, z=~var3,
        color=~var4, colors=c("royalblue", "coral") ) %>%  add_
markers()
```

The result is shown in the following figure.

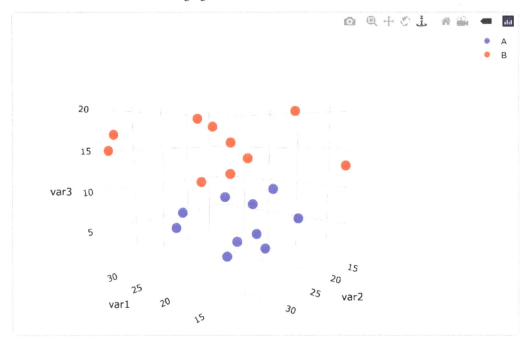

Figure 11.14 – 3D scatterplot of two similar variables. The separation occurs in the third dimension

Looking at *Figure 11.14*, it becomes clear the separation when the third dimension `var3` is added to the plot. The plot has changed from aligned points to more scattered points.

These are two cases where the use of 3D plots is recommended. Other than that, it is not the best option because the third axis makes the graphic more complex, and the more points are added to the dataset, the more difficult it will be to interpret it. Additionally, 3D graphics are not good when precision is required in a comparison. If you look at *Figure 11.14* and *Figure 11.15* and try to determine the exact numbers where the points are located, that task will be more difficult with the naked eye, since our brain captures the real world in two dimensions, but it simulates a third dimension on a flat surface by comparing the images from both eyes (`https://tinyurl.com/mr38amt4`). That calculation can be complex, making our interpretation of a 3D plot imprecise.

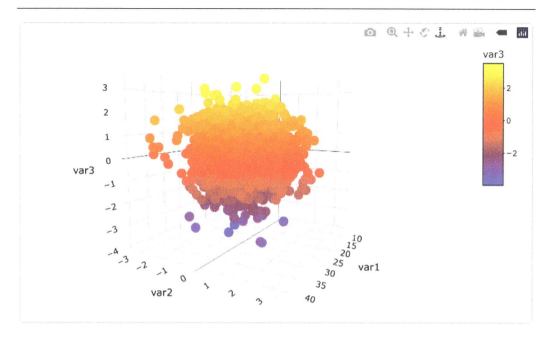

Figure 11.15 – Busy 3D scatterplot that is difficult to read

Look at the graphic in *Figure 11.15*. There are so many points that it is too complex for us to process and understand what is going on.

Next, we will learn about adding interactivity to a plot in RStudio.

Adding interactivity to graphics

Images are interpreted by our brains faster than words or numbers (`https://tinyurl.com/ nhtbw9jk`). That makes graphics an interesting way to show data, as we have learned throughout this book. But there is still more enhancement to be done when working with data visualization, and one of these enhancements is interactivity.

The `ggplot2` library creates static graphics. Hence, the plots will not show values at the tops of bars or names of points on a scatterplot, for example. If that is a requirement for a visualization, it must be added using an annotation or text. However, when you combine the graphic's code with **plotly**, some interaction is added to the visualization, such as making values appear just by hovering over a data point or zooming in and out the graphic.

To create an interactive scatterplot out of the same code that generated in *Figure 11.1*, we only have to add the `ggplotly()` function around the entire `ggplot` code. See the following code snippet:

```
# Scatterplot of price by carat.
ggplotly(
  ggplot(df) +
  geom_point( aes(x= carat, y= price, color= cut,
alpha=0.5))  )
```

The resulting graphic is similar to the one in *Figure 11.1*, but this one has some interactive enhancements.

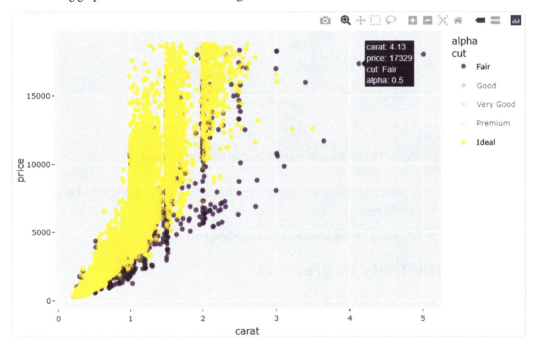

Figure 11.16 – Interactive plot: a combination of ggplot2 and plotly

Figure 11.16 displays a few buttons, such as snapshot, zoom, and custom selection, as well as the ability to click on the legend names to show or hide a group (in this graphic, only **Fair** and **Ideal** cuts are selected). Additionally, when we hover over a point, it shows a balloon with the numbers associated with it, as shown for one of the points in the top right-hand corner.

This is the end of this chapter. We are now able to create basic enhanced visualizations. From this point on, to make your plots even better, go to the documentation pages and study the functions, take advantage of the example code snippets they provide, and create your own enhanced plots.

In the next chapter, we will learn about two other visualization options: using R code to draw plots in Microsoft Power BI, and word clouds.

Summary

After reading this chapter, you should be able to make enhanced plots, such as facet grids, maps, and 3D plots.

We started by learning about facet grids, which are one of the grammatical elements of the grammar of graphics. With facet grids, a graphic can be divided into subplots, making the interpretation easier for the reader. The next topics were how to plot maps and time series in R using ggplot2. These are vast subjects that lie within geospatial data analysis and time series analysis in data science, so we just covered the basics, but that should be enough for you to create great visualizations.

3-dimensional plots are beautiful and impactful, no doubt. However, they are not well suited for big data or for visualizations where precision is a requirement. They are good, though, for plotting surfaces or viewing the separation of data points that is only visible with the addition of a third dimension.

Finally, we closed the chapter with a function that combines `plotly` and `ggplot2`, bringing interactivity to our visualization.

Exercises

1. What is a facet grid and what is the function used to create it using `ggplot2`?
2. What step is required before plotting a map with `ggplot2`?
3. What is the geometry used in order to create maps on `ggplot2`?
4. What library is used to create 3D plots?
5. What characterizes a dataset as a time series?
6. List two use cases where 3D plots can be recommended.
7. What is the function from `plotly` for adding interactivity to `ggplot2` graphics?

Further reading

- About the diamonds dataset: Type `?diamonds` in the RStudio console
- `ggplot2` cheatsheet: `https://github.com/rstudio/cheatsheets/blob/main/data-visualization.pdf`
- Geospatial maps with ggplot2: `https://ggplot2-book.org/maps.html`
- Get started with tmap, another library for geospatial data plotting: `https://tinyurl.com/2p9b9522`
- `plotly` for R graphic gallery: `https://plotly.com/r/`
- Time series with ggplot2: `https://r-graph-gallery.com/279-plotting-time-series-with-ggplot2.html`
- 3D charts with plotly: `https://plotly.com/r/3d-charts/`
- R code in GitHub for this chapter: `https://tinyurl.com/yckwhy64`

12

Other Data Visualization Options

Data visualization is an important part of data science. There are a lot of resources available, as we have been seeing throughout *Part 3* of the book. In this last chapter of *Part 3*, we will go over two extra visualization options:

- Plotting graphics in Microsoft Power BI using R
- Preparing data for plotting
- Creating word clouds in RStudio

Technical requirements

All the code can be found in the book's GitHub repository: `https://github.com/PacktPublishing/Data-Wrangling-with-R/tree/main/Part3/Chapter12`.

The following are the libraries to load to RStudio for this chapter:

```
library(tidyverse)
library(wordcloud2)
library(officer)
library(tidytext)
```

Plotting graphics in Microsoft Power BI using R

Many **business intelligence** (**BI**) tools, such as **Microsoft Power BI**, have been developed and launched in the last few years. Microsoft's tool appeared in 2014 and it currently accepts its own graphics and integrates with programming languages such as R and Python.

Working with BI tools is very practical. Most of what can be done in terms of visualization does not require coding; instead, you can just drag and drop variables to create graphics. However, like any other tool, there are pros and cons. For example, if we want to create a more customized graphic or add a type of graphic that is not available in Power BI, such as a histogram, we will have to adapt and look for an alternative to create that visualization. For that reason, we will learn how to plot a **ggplot2** graphic in Power BI. It will give us the flexibility to make much more than the standard graphics provided by the BI tool.

Let's see how to plot histograms in Power BI using R code.

The first step is to have the dataset in the BI tool. It can be any dataset, so for this purpose, we will generate some random data in R and load it into **Power BI Desktop** (the **Home** tab > **Get Data** > **Text/CSV**). Here is an extract of the dataset created.

	var1	var2
1	1.325697190	0.4549387111
2	0.126779955	0.0817962112
3	-0.300655183	0.2339162868
4	-0.634663787	0.0009492298

Figure 12.1 – Random data generated and loaded to Power BI

After we have opened the software and loaded the dataset to our session, click on the ⬛**R** button on the visualizations side panel. The **Enable script visuals** pop-up shown in *Figure 12.2* will appear. Click on **Enable**.

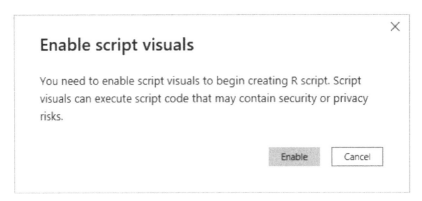

Figure 12.2 – Enable R script visuals in Power BI

Then, your screen will look something like *Figure 12.3*.

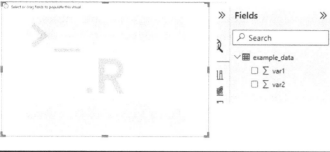

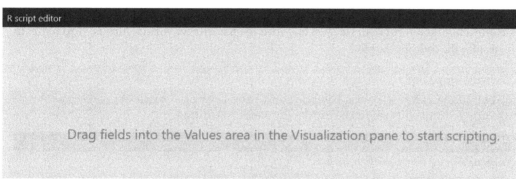

Figure 12.3 – Power BI screen after insert a R

As seen in *Figure 12.3*, now we can drag the variables we want to work with to the plot area (top left rectangle). Since we want to create a histogram, we will only drag var1. Once we do that, notice that Power BI starts to populate some comments with R code in the gray area at the bottom of the screen.

R script editor

⚠ **Duplicate rows will be removed from the data.**

```
1  # The following code to create a dataframe an
2
3  # dataset <- data.frame(var1)
4  # dataset <- unique(dataset)
5
6  # Paste or type your script code here:
```

Figure 12.4 – R code being populated by Power BI

A few notes before we proceed. The software uses the `dataset` variable name that holds the `data.frame` object. Therefore, all your code must be written with that in mind. Power BI creates a data frame with all the variables you drop on the plotting area. It also removes duplicated entries. Also, the code is created in the background, but it shows as commented for your information; thus, you should not remove the # from it, as you will get an error.

Furthermore, as this feature is an integration between two different pieces of software, it may not always work as expected due to dependencies or new version releases from one side or the other. At the time of writing, the most recent supported version of R in Power BI was 3.4.4 (`https://tinyurl.com/63u2476w`).

To continue, notice the message saying we should paste (or write) our script to create the graphic. Let's create a histogram. For that, we must also load the libraries we will use, that is, `ggplot2`, in this case, plus the code for the plot:

```
library(ggplot2)
ggplot(dataset) + geom_histogram( aes(var1), bins=8, color=
"white", fill = "royalblue") + theme_classic()
```

R script editor

⚠ Duplicate rows will be removed from the data.

```
1  # The following code to create a dataframe and remove duplicated rows is always executed and acts as a preamle for your script:
2
3  # dataset <- data.frame(varl, var2)
4  # dataset <- unique(dataset)
5
6  # Paste or type your script code here:
7  library(ggplot2)

8  ggplot(dataset) + geon_histogram( aes(var1), buns-8, color-"white", fill-"royalblue"royalblue" | + theme_classic( )
```

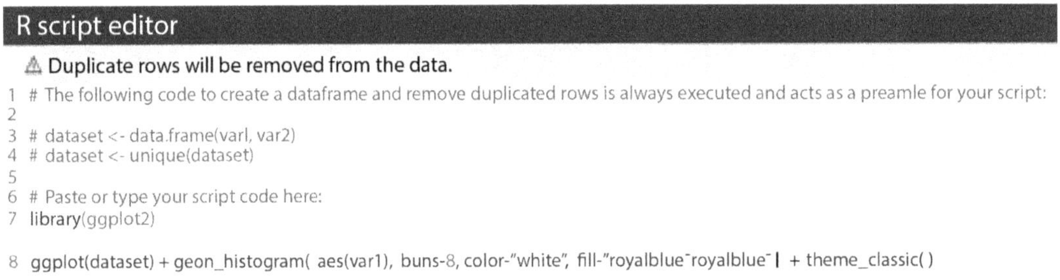

Figure 12.5 – R code for a histogram in Power BI

The result will be as shown in the next figure.

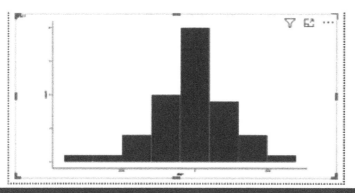

⚠ Duplicate rows will be removed from the data.

```
1
2
3  # dataset <- data.frame(varl, var2)
4  # dataset <- unique(dataset)
5
6  # Paste or type your script code here:
7  library(ggplot2)

8  ggplot(dataset) + geon_histogram( aes(var1), buns-8, color-"white", fill-"royalblue¬royalblue¬| + theme_classic( )
```

Figure 12.6 – R code and the resulting histogram in Power BI

Figure 12.6 shows the result of the R code running within Power BI Desktop. The resulting graphic is the same as we get in RStudio but embedded in another application. The advantage of knowing how to create these plots is adding even more power to the BI tools and opening up room for more customized dashboards.

In the sequence, we will study how to prepare a dataset for plotting. After all, each graphic has different inputs. Sometimes, data must be wrangled to fit the right format for a plot. The example we will use is a word cloud.

Preparing data for plotting

There are many kinds of graphics, for instance, univariate, bivariate with one numeric and one categorical variable or two numeric variables, and others. The input data will be different for each of them, requiring the data scientist to munge the data to fit a specific format to plot it. An example studied in this book was the data that was not in tidy format, requiring transformations before it could be plotted.

In this section, we will learn how to prepare a text to be plotted as a word cloud, which is a graphical way to show the content of text. A word cloud is a graphical representation of the most frequent words that appear in a text. The more frequently the word occurs, the bigger it appears in the plot, consequently providing a sense of the content of the text.

A text is a combination of words, but it does not have rows and columns of data. Instead, it is a whole piece. So, prior to plotting the word cloud, it is necessary to transform the data to the correct input format required by the `wordcloud2` library. This library expects to receive a data frame or tibble object with the words and the frequency they appear in the text, so let's move forward and transform a text in a tibble object.

We will use the text from *Chapter 10* as the input chosen to create the word cloud and to look at the most frequent words. Here are the libraries used for this exercise:

```
library(tidyverse)
library(wordcloud2)
library(officer)
library(tidytext)
```

We begin by reading the word document. We use the officer library to read `.docx` files. The file will be read and stored in a variable named `chapter10`. This is a list object, which, in the R language, is a group of objects of multiple types. But we are only interested in the text, so we are inputting `chapter10` in the `docx_summary()` function, which returns a `data.frame` object:

```
# Read word document
chapter10 <- read_docx("Data_Wrangling_With_R_Chapter10.docx")
content <- docx_summary(chapter10)
```

The data frame in the variable content is shown next, in *Figure 12.7*.

	doc_index	content_type	style_name	text
1	1	paragraph	H1 - Chapter	Chapter 10
2	2	paragraph	H1 - Chapter	Introduction to ggplot2
3	3	paragraph	P - Regular	
4	4	paragraph	P - Regular	
5	5	paragraph	P - Regular	In the part 3 of this book, beginning now, the subject is dat...

Figure 12.7 – Data frame resulting from the .docx document

Moving forward, we create a variable named `text` to get the `content` data frame, select only the `text` column, which is our object of interest, transform the blank cells to `NA` and then drop those cells, given that we want only text:

```
# Extract only textual information and drop Blank cells and
NAs.
```

```
text <- content %>% select(text) %>%
  na_if("") %>% drop_na()
```

As we will use the `tidytext` library for the next transformation, we will convert the text variable to a tibble. Tidy libraries work much better with that type of object:

```
# Transform to tibble object
text <- tibble(text)
```

The following step is to tokenize the text. This task is very common in the **natural language processing (NLP)** area, and it means breaking down the text into a minimal unit that carries a meaning, usually a word. Therefore, one token, here, means one word. This can be done using the `unnest_tokens()` function, which needs as arguments the name of the resulting `output=word` column, and the column to be used as input goes in the argument with the same name, `input= text`. The result is stored in the `text_tokens` variable:

```
# Tokenize - one word is one token
text_tokens <- text %>%
  unnest_tokens(output= word, input= text)
```

The result will be a single-column tibble, as seen in *Figure 12.8*.

	word
1	chapter
2	10
3	introduction
4	to
5	ggplot2
6	in

Figure 12.8 – Tibble with the tokenized text

Observe that the `unnest_tokens()` function does other important transformations that are very common in text mining, such as converting the tokens to lowercase, removing text punctuation, and if there were the line numbers where the word came from, these would also be removed. So, it is a complete function that does a lot for us.

These two code snippets are needed in the sequence to remove numbers and other eventual punctuation from our dataset, such as decimal numbers. The first code snippet receives `text_tokens` and filters the column word using the `str_detect()` function from `stringr`, and the regexp `"\\D"`, which means no digits. The result is stored in `clean_tokens`:

```
# Remove numbers
clean_tokens <- text_tokens %>%
  filter(str_detect(word, "\\D"))
```

Then, we get `clean_tokens` and filter for results different from punctuation. That filter expression is created with the `!` signal (that means NOT) and the regexp for punctuation:

```
# Remove punctuation
clean_tokens <- clean_tokens %>%
  filter(!str_detect(word, "[:punct:]"))
```

Continuing the cleaning, now it is time to remove *stop words*. These are words that do not carry any meaning for the text, such as conjunctions, prepositions, and pronouns (the, it, in, or, and so on). This code brings the `anti_join(stop_words)` function that keeps only the observations that are not contained in the `stop_words` object. Once again, we overwrite `clean_tokens`:

```
# load stopwords
data(stop_words)
# Anti-join: keep only what is not in stop words
clean_tokens <- clean_tokens %>%
   anti_join(stop_words)
```

We finally have the clean dataset. The next step is to add a column with the counts of each word and store it in a new variable called `word_freq`. Here is the code to do that:

```
# Count word frequency.
word_freq <- clean_tokens %>%
   count(word, sort = TRUE)
```

After all these cleanup codes, the resultant tibble object is displayed in *Figure 12.9*.

	word	n
1	plot	46
2	graphic	44
3	figure	31
4	ggplot2	29
5	color	27

Figure 12.9 – Dataset ready for input in the word cloud

The resultant dataset is the input needed to create a word cloud.

The preparation of the data for plotting is an important part of data wrangling. To perform the preparation, before starting, we must know the exact format needed by the graphic library that will receive it. As mentioned at the beginning of this section, `wordcloud2` requires a data frame with the words and the respective frequencies to plot the visualization; ergo, we worked all the steps toward that goal. That was the lesson we wanted you to learn in this section.

Let's use the clean dataset in the sequence to create the word cloud.

Creating word clouds in RStudio

A word cloud is very useful when we want to visualize the content of a text quickly. The more times a word is repeated within the text, the bigger it will be displayed on the word cloud, giving us a sense of what we can expect if we read the text. It is kind of a summary.

Once we have a dataset with words and their frequencies, plotting a word cloud takes just one line of code. The function is the same name as the library, `wordcloud2()`, and it takes as inputs the dataset, a color palette or a vector of colors, and the size of the words:

```
# Generating WordCloud
wordcloud2(data=word_freq, color="random-dark", size=1)
```

The result after running this code is printed as follows.

Figure 12.10 – Word cloud generated with the content from Chapter 10 of this book

Just to refresh our minds, *Chapter 10* is about an introduction to the `ggplot2` library. It brings the concepts of the grammar of graphics and introduces the syntax for many kinds of graphics using the mentioned library.

Now, looking at *Figure 12.10*, it summarizes the content of the chapter very well. Observe that the largest words are those we talk a lot about: *plot*, *figure*, *colors*, *aesthetics*, and *ggplot2*. We can also notice graphical elements, such as *bar*, *line*, *histogram*, *fill*, *aes*, and *axes*. Furthermore, since we used the *mtcars* dataset for many examples, the words such as *mpg*, *gallon*, and *miles* also pop up.

When we look at a word cloud, we should get a feel for the text's content. That task was fulfilled in this case.

This word cloud marks the end of the chapter and of the third part of this book. The sequence is the last part, where we talk about modeling and deploying. In the next two chapters, we will consolidate our learnings throughout the entire book and create a model that will later be deployed in an application built with Shiny.

Summary

In this brief chapter, we covered some additional options for visualization. We began the chapter by showing how we can integrate ggplot2 graphics in Microsoft Power BI, enhancing the capabilities of the tool. Next, we moved on to learn in practice how we can prepare data for plotting, with the creation of word clouds as the final goal, and, at the end of the chapter, we learned how to plot one and how to interpret it.

Exercises

1. What programming languages can integrate with Power BI?

2. What is the benefit of plotting R graphics in Power BI?

3. What should you have in mind when preparing data for plotting?

4. List the library name and function to create a word cloud.

5. What are the two variables needed in the input dataset for a word cloud?

Further reading

- Text mining with R: `https://www.tidytextmining.com/tidytext.html`

- Wordcloud2 introduction from CRAN: `https://tinyurl.com/3rvnatkc`

- Graphic library with code snippets for wordcloud2: `https://tinyurl.com/2vkpa2tw`

- Microsoft documentation: *Create Power BI visuals using R*: `https://tinyurl.com/mry85twv`

- *R Visuals in Power BI – Histogram*: `https://youtu.be/nePWIVgHobs`

- R code on GitHub for this chapter: `https://tinyurl.com/54kjhcwk`

Part 4: Modeling

This part includes the following chapters:

13

Building a Model with R

The last part of this book is about modeling data. It has been a long learning journey so far. We started with the fundamentals of data wrangling while covering the concepts that surround the matter and going through techniques to munge each type of data. During practical projects, we had the opportunity to wrangle entire datasets, showing some transformations. In the previous part, we worked with plenty of material regarding data visualization while going over one of the most complete libraries for visualization.

Now, it is time to put all our knowledge to the test and work on a final project. This will involve end-to-end work, from loading the dataset into RStudio to deploying it in a production environment using **Shiny**, where anyone can interact with the application.

This project will be built during these last two chapters in *Part 4*. Let's get to work.

We will be covering the following main topics:

- Machine learning concepts
- Understanding the project
- Preparing data for modeling in R
- Exploring the data with a few visualizations
- Selecting the best variables
- Modeling

Technical requirements

The dataset to be used in this project is called *Spambase* and is from the UCI Machine Learning Repository (https://archive.ics.uci.edu/ml/datasets/spambase).

The code for this chapter can be found in this book's GitHub repository: https://github.com/PacktPublishing/Data-Wrangling-with-R/tree/main/Part4/Chapter13.

The following libraries will be used in the code:

```
library(tidyverse)
library(patchwork)
library(skimr)
library(randomForest)
library(caret)
library(ROCR)
```

Machine learning concepts

Before we move on to the project itself, let's just build a background about machine learning concepts. This content is not the main scope of this book; therefore, we will quickly go over a couple of definitions to put us on the same page for the remainder of this book.

A model is a representation of a theory (HAIR Jr. et al, 2019) but is also defined as a simplification or approximation of reality (Burnham & Anderson, 2002). In other words, modeling data involves finding patterns that can help us explain a response, which is the most probable outcome from that observation.

With that said, the model will just reflect the data that it received. For that reason, it is crucial that the input data is clean and representative of the reality we are trying to model. To exemplify this, think about when we see a dataset with too many missing values that are going to be either removed or inputted. Both approaches will certainly have an impact on the model, so it must be done carefully, as eliminating an important variable, as well as imputing too many values, can distort the result.

In machine learning, there are classification and regression models. Both models will be described next.

Classification models

These models are used for categorical output. So, let's say we have a lot of medical data listing symptoms, a person's lifestyle, and physical characteristics and we want to look at the data to find patterns that can lead us to predict if a given patient has a higher chance of having a heart attack, for example. This is a classification because we are transforming the input data into patterns that will be separated into groups: a high chance to have a heart attack or a low chance to have a heart attack. Thus, each observation will be classified into a group, according to its patterns. Examples of classification algorithms include KNN, decision tree, random forest, logistic regression, and support vector machine.

Regression models

Regression models, on the other hand, return numerical output. They find patterns in numbers and return a continuous output such as a price, salary, or age of a person. A classic problem that can be solved with regression models is predicting the prices of houses, where the input is the construction area and the number of bathrooms and bedrooms, and it returns an estimated price for the house. We can assess that these models make sense if we think about how the housing market works. Consider two houses in the same neighborhood that are similar in size and have the same number of bedrooms and bathrooms. Those two houses will likely be evaluated within the same price range. So, that is what the regression algorithm will do – it will come up with an equation that captures the relationship between the input variables and calculate an estimate of the output. Examples of regression algorithms include linear regression, polynomial regression, and regression tree.

Supervised and unsupervised learning

Furthermore, in terms of definitions, there are three major types of machine learning algorithms:

- **Supervised**: This type of algorithm receives data containing variables that can explain an outcome. These models work just like a student that is studying for a test. You provide the content and the answers to *teach* the algorithm how to solve the problem using those variables. That is the training part of the process. Once the algorithm has studied enough and is getting *good grades* in the exercises, you apply a test, giving the algorithm some new data it's never seen before to see if it can generalize the solution to unseen data. Notice in *Figure 13.1* that the output is the same when the input is the same. This pattern would easily be identified by a supervised learning algorithm. Examples of where this algorithm can be used include classification and regression models.

- **Unsupervised**: This type of algorithm does not receive a labeled variable – it just reads the dataset, looking for patterns that can help us explain the data. *Figure 13.1* shows some data points where similar colors are clustered together. An unsupervised learning algorithm can capture that pattern and put similar colors within the same buckets so that they can be analyzed in depth later. Those points could be representing clients that shop in different locations of a retail company. Clustering models use these types of algorithms.

- **Reinforcement**: These algorithms learn by trial and error. They start by performing an action and checking if the outcome is going in the right direction. If it's positive, then the algorithm is rewarded; otherwise, it is penalized. The objective of these algorithms is to reduce the penalties to the minimum possible. Observe *Figure 13.1* once again to see that the algorithm makes a couple of corrections while going from point A to B. Whenever it understands that the calculations are too high or too low, the path is corrected. Video games use these types of algorithms:

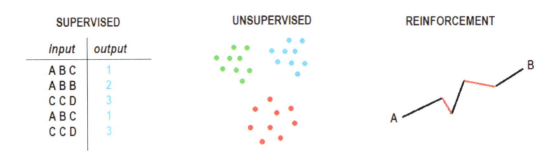

Figure 13.1 – Learning types for algorithms

A model will always only be as good as the quality of the data that is fed to it. If the input isn't of high enough quality, the project should be returned to the data-gathering phase.

Next, we will understand the project.

Understanding the project

When starting a project, we need a purpose – that is, a goal we want to reach at the end. After all, knowing the problem is part of the solution. Like Lewis Carrol wrote in his book *Alice's Adventures in Wonderland*, the Bunny says to Alice that if she does not know where she wants to go, any path will lead her there.

So, let's begin by understanding the project, or where we want to go.

The dataset

The input data for this project is the *Spambase Data Set* (`https://tinyurl.com/23xwdcah`), which can be found in the UCI datasets repository. See the citation information in the *Further reading* section at the end of this chapter for more.

It contains 4,601 observations and 57 explanatory variables. Out of those, 48 features are floating numbers representing the percentage value, from 0 to 100, of specific words associated with spam and their percentage present in the message. There are six other variables with special characters such as parentheses or dollar signs and their percentage present in the message, and three with integer numbers representing the number of uppercase letters present in the message. Additionally, there is one target variable, which is a label classification of *spam* (represented by the number 1) or *not spam* (0) that we will predict.

As per the creators, this data was created by the postmaster and from individuals who marked messages as spam in their inboxes in 1999.

The project

As defined by the Merriam-Webster dictionary, *"a spam email is an unsolicited message sent to a large number of recipients."* Those messages can be a problem for many email providers, as well as an inconvenience and a security risk for the person who receives them, given that many of those emails also carry malicious content that explores security breaches in the computers or mobile devices connected to the internet. Hence, nowadays, every email client is equipped with spam detectors, with many of those being operated by **artificial intelligence (AI)** algorithms.

There are many ways to create a spam detector using AI models, so we will explore one of them in this project.

Imagine that we are a company that works with advertising and sends a lot of commercial emails to many clients. That is a potential spam generator. So, to decrease the number of messages that end up in the spam filter, our client has asked us to create a spam classifier capable of reading a message and predicting the probability of it being marked as spam or not. The client has a dataset containing some words. It specifies how many are present in the text (in percent) and if the email that contains those words was marked as spam or not.

Our goal, therefore, is to create a tool that can get any text as input and, based on a trained classification algorithm decision, estimates (or predicts, if you will) the probability of that message being classified as spam or not by a spam detector, increasing the company's chances that the marketing message will reach their final clients.

The algorithm

This is a classification project where we must return the probability of each label: whether the email is *spam* or *not spam*. We will use the **random forest** algorithm for this. It is a widely known classification algorithm that uses an ensemble learning method – more specifically, **bootstrap aggregating**. Simply put, such a method extracts a sample of the dataset and generates a decision tree. The tree goes over each variable, creating true or false statements based on the data and calculating a measurement of information gain for each of those decisions, helping the algorithm to move on to the next decision for another variable. That stream of decisions for each variable creates paths that, when combined, will look like an inverted tree (hence the name). The algorithm keeps going until it reaches a classification, such as spam or not spam.

Figure 13.2 shows an example of a simple decision tree for us to build a better intuition about the concept. Starting from the root question *Is it raining?*, based on each answer, we move forward and create another branch. If it is not raining, we can go out. If it is, we ask ourselves whether or not we have an umbrella. If so, we go out; otherwise, we check if we have a raincoat. If so, we are good to go out. If not, we will stay home. Observe how we moved down from the root question to a final decision on whether to go out or not. Such decisions are called **leaves** on decision trees, and the algorithm works just like that, creating binary questions to reach the leaf level for each observation in a dataset:

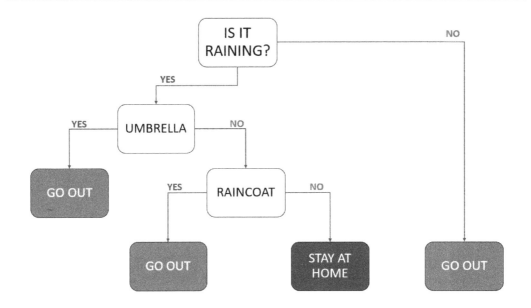

Figure 13.2 – Example of a simple decision tree

In ensemble methods such as **random forest**, hundreds of decision trees are created out of different samples of the dataset. The final classification of each observation will be based on an average of all the answers from the trees created.

Now that we know our problem and that we will use a **random forest** classification model to solve it, let's load the dataset into RStudio and start working.

Preparing data for modeling in R

We must wrangle the data to prepare it for modeling. Since we know where we want to go at the end of this project, the next step is a matter of finding a way to get there.

The first thing we must do is load the libraries to be used for wrangling and modeling the data. We will use **tidyverse** to perform data wrangling and visualization, **skimr** to create a descriptive statistics summary, **patchwork**, a great library to put graphics side by side, **randomForest** to create the model, **caret** to create the confusion matrix, and **ROCR** to plot the ROC curve of model performance.

To load the dataset, the best option is to pull it directly from the internet, without the need to save it locally on our machine. Just use the read_csv() function and point to the web address where the raw dataset is located, as we've done previously in this book. Here, we are using the trim_ws=TRUE argument to trim any unwanted white spaces and the col_names=headers argument, where headers is a vector that's created with all the variables' names. The header names can be found in the dataset documentation in UCI or the code for this chapter, which can be found on GitHub (https://tinyurl.com/mr497d8v):

```
# Link where the dataset is located in UCI database
url <- "https://archive.ics.uci.edu/ml/machine-learning-
databases/spambase/spambase.data"
# Load the dataset
spam <- read_csv(url, col_names = headers, trim_ws = TRUE)
```

An extract of the *spam* dataset can be seen in *Figure 13.3*:

	word_freq_make	word_freq_address	word_freq_all	word_freq_3d	word_freq_our
1	0.00	0.64	0.64	0	0.32
2	0.21	0.28	0.50	0	0.14
3	0.06	0.00	0.71	0	1.23
4	0.00	0.00	0.00	0	0.63
5	0.00	0.00	0.00	0	0.63

Figure 13.3 – Spambase dataset from the UCI dataset repository

Once the data has been loaded, we must understand the variables and the data types so that we know how to proceed when we are wrangling the data. Remember that each data type is suited for certain kinds of transformations. As we have seen, the transformations we can make for text are not the same as those applied to numbers.

To begin understanding the dataset, the best function to use is glimpse() as it will give us its dimensions, as well as its data types. Look at the first few rows of the output to find the number of rows and columns, as well as the variable type between <>:

```
# Glimpse of the data
glimpse(spam)
```

```
[1] 4601    58
Rows: 4,601
Columns: 58
$ word_freq_make        <dbl> 0.00, 0.21, 0.06, 0.00, 0.00, 0.00, 0.00, 0.00, 0.15, 0.~
$ word_freq_address     <dbl> 0.64, 0.28, 0.00, 0.00, 0.00, 0.00, 0.00, 0.00, 0.00, 0.~
$ word_freq_all         <dbl> 0.64, 0.50, 0.71, 0.00, 0.00, 0.00, 0.00, 0.00, 0.46, 0.~
$ word_freq_3d          <dbl> 0, 0, 0, 0, 0, 0, 0, 0, 0, 0, 0, 0, 0, 0, 0, 0, 0, 0,~
$ word_freq_our         <dbl> 0.32, 0.14, 1.23, 0.63, 0.63, 1.85, 1.92, 1.88, 0.61, 0.~
$ word_freq_over        <dbl> 0.00, 0.28, 0.19, 0.00, 0.00, 0.00, 0.00, 0.00, 0.00, 0.~
$ word_freq_remove      <dbl> 0.00, 0.21, 0.19, 0.31, 0.31, 0.00, 0.00, 0.00, 0.30, 0.~
$ word_freq_internet    <dbl> 0.00, 0.07, 0.12, 0.63, 0.63, 1.85, 0.00, 1.88, 0.00, 0.~
```

Figure 13.4 – Glimpse of the spam dataset

Figure 13.4 shows some of the variables. There are 57, so we can't see them all here, but they are mostly from the double type. The last three, which are related to how many uppercase letters are in the text, are integers and can be converted. The same goes for the target variable, spam, which should be converted into a factor. We will do that with the following code, where the cols_to_int and cols_to_factor variables are collecting the names of the columns to be converted into other data types. Then, we can use the mutate_at() function to make the transformations:

```
# Columns to change to factor
cols_to_int <- c("capital_run_length_average", "capital_run_length_longest", "capital_run_length_total")
cols_to_factor <- c("spam")


# Assign variables as factor
spam <- spam %>%
  mutate_at(cols_to_int, as.integer) %>%
  mutate_at(cols_to_factor, factor)
```

The following check verifies the presence of missing values:

```
# Check for NA
sum( is.na(spam) )
```

This is the output of the check for missing values:

```
[1]  0
```

We called for a sum of the values that are NA. As seen in the output, there are no missing values, so we can move on with the exploration. So, let's call the skim(spam) function to pull the descriptive statistics and start to get a sense of how spread out the data is and what the distributions of the variables are. Before using the skim function, we will adjust the display options to show floating numbers instead of scientific notation:

```
# Configure display not scientific numbers
options(scipen = 999, digits=3)


# Descriptive statistics
skim(spam)
```

The result can be seen in *Figure 13.5*, which we are already familiar with at this point:

	skin_variable <chr>	n_missing <int>	complete_rate <dbl>	ordered <lgl>	n_unique <int>	top_counts <chr>
1	spam	0	1	FALSE	2	0: 2788, 1: 1813

	skin_variable <chr>	n_missing <int>	complete_rate <dbl>	mean <dbl>	sd <dbl>	p0 <dbl>	p25 <dbl>	p50 <dbl>
1	word_freq_make	0	1	0.10455	0.3054	0	0.00	0.000
2	word_freq_address	0	1	0.21301	1.2906	0	0.00	0.000
3	word_freq_all	0	1	0.28066	0.5041	0	0.00	0.000
4	word_freq_3d	0	1	0.06542	1.3952	0	0.00	0.000
5	word_freq_our	0	1	0.31222	0.6725	0	0.00	0.000

Figure 13.5 – Descriptive statistics from the spam database

The data displayed in *Figure 13.5* is a partial screenshot. But looking at the whole table, a few observations can be made:

- There are many means close to zero, so most of the values never go too high.

- There are some variables where more than 75% of the observations are zeros.

- The standard deviations are high, denoting the presence of outliers or highly spread data.

- The histograms may be skewed to the right.

- There are 2,788 (approximately 60%) non-spam messages and 1,813 (approximately 40%) spam messages. The data is close to balanced.

The importance of bringing descriptive statistics in is that we can start gaining insights about the data's behavior. From the points previously listed, we can infer that the model will probably not give too much importance to the majority of the words present in a text, given that they appear just a few times, which is not enough to create a pattern to classify a text as spam.

We have a good understanding of the dataset at this point. Now, it is time to create a few visualizations so that we can dive deeper and prepare ourselves for data modeling.

Exploring the data with a few visualizations

We should start the data visualization portion of a project with univariate graphics, such as histograms and boxplots. This is because the former will show us the data distribution, indicating the possible statistical tests to be used, while the latter will bring up the presence of outliers in the data.

Since there are more than 50 variables in this dataset, we will create a `for` loop to plot the histograms for all of them. The following code uses the `hist()` function from the base R histogram:

```
# Histograms
for (var in colnames(spam)[1:57]) {
```

```
hist(unlist(spam[,var]), col="royalblue",
     main= paste("Histogram of", var),
     xlab=var)    }
```

Notice that we only did the loop for columns [1:57] since we know that the last one is the target variable. Next, we will see four graphics, as shown in *Figure 13.6*:

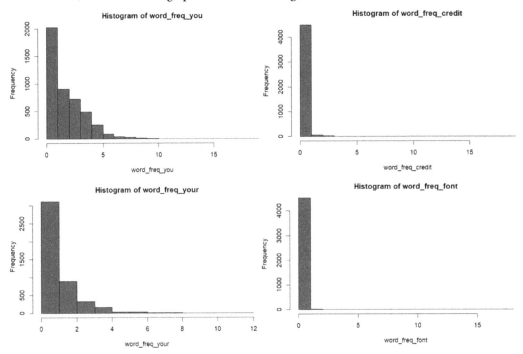

Figure 13.6 – Histograms from the Spambase dataset

Look at how the distributions are concentrated around the zero value and very skewed to the right. These plots confirm that the descriptive statistics make sense.

In the sequence, we will compare the boxplots of the variables so that we understand which of them are impacting the message so that it's classified as spam. However, if we take another look at the table in *Figure 13.3*, we will see that it is not Tidy formatted.

As a reminder, a Tidy dataset needs to have one variable per column, which is partially true in this case.

We can see that the columns are words, so, in this case, they should be under the same variable, word, while the numbers should go on a separate column. For **ggplot2**, this is the needed format for plotting, so we must transform the dataset from a wide format into a long format, placing the columns under the same single variable.

In the following code, we are using `pivot_longer()` to do this, providing it with the columns to be pivoted, the variable name to receive the column names, the words (`names_to`), and the variable name to receive the values (`values_to`). The result is stored in `long_spam`:

```
# Pivot the table to Long format
long_spam <- spam %>%
  pivot_longer(cols=1:57, names_to= "words", values_to = "pct")
```

The result will look as follows:

spam	words	pct
1	word_freq_make	0
1	word_freq_address	0.64
1	word_freq_all	0.64
1	word_freq_3d	0
1	word_freq_our	0.32
1	word_freq_over	0
1	word_freq_remove	0
1	word_freq_internet	0
1	word_freq_order	0
1	word_freq_mail	0

Figure 13.7 – Spambase dataset in Tidy format

With this transformation, it is ready for plotting. The boxplots will help us answer whether words are appearing more frequently than others, hence impacting a future classification, and which words have more outliers, if any, that can distort our classifier. The following code snippet takes the `long_spam` dataset, filtering only the observations starting with `word_` that were classified as spam (`spam == 1`). Then, we are creating a boxplot and adding a title, subtitle, and theme. As a reminder, `reorder(words,pct)` is used to arrange the names on the *Y*-axis by the percentage, `pct`:

```
# Top 10 plot
long_spam %>%
  filter(str_detect(words,"word_") & spam == 1) %>%
  ggplot()+
  geom_boxplot( aes(y=reorder(words,pct), x=pct, fill=spam))+
  ggtitle("Percentages of words and their association with spam
e-mails") +
  labs(subtitle= "The frequency of appearance of some words in
e-mails is more associated with spam.") +
```

```
labs(x= "Percentage", y= "Word") +
theme_classic() +
theme(plot.subtitle = element_text(color = "darkgray",
size=11) )
```

The original plot isn't the best for printing, given the large number of variables. See *Figure 13.8*:

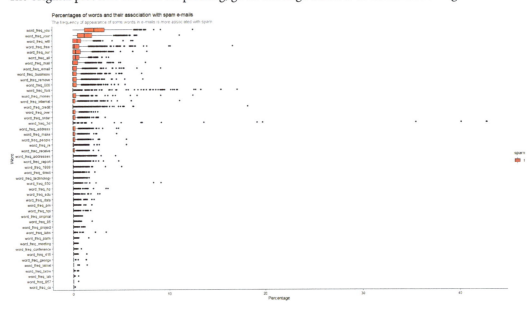

Figure 13.8 – Top 20 words associated with spam emails

In this case, I will display another graphic that shows only the top 20 words that are more associated with spam emails so we have a better idea:

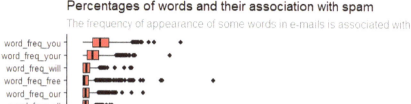

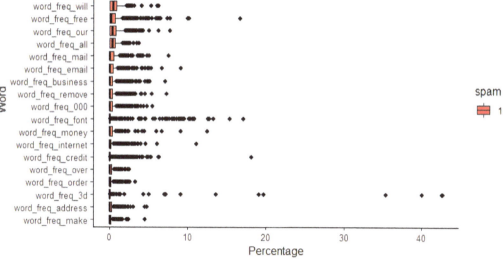

Figure 13.9 – Top 20 words associated with spam emails

Looking at *Figure 13.8* and *Figure 13.9*, we can that after the 24th record or so (more or less in the middle of the *Y*-axis), all the boxplots have their medians too close to zero, so they don't impact the spam classification too much.

Once we have determined that the top 23 words are more relevant and can have more impact on a classification, we must plot the boxplots while comparing observations marked as spam (1) and not spam (0) from the dataset to see how they impact the top 23 words and the entire data.

We will start by creating a vector to gather the top 23 words:

```
# Define top words
top_words <- c("word_freq_you", "word_freq_your", "word_freq_
will", "word_freq_free","word_freq_our", "word_freq_all",
"word_freq_mail", "word_freq_email", "word_freq_business",
"word_freq_remove", "word_freq_000", "word_freq_font","word_
freq_money", word_freq_internet", "word_freq_credit","word_
freq_over",                  "word_freq_order", "word_freq_3d",
"word_freq_address", "word_freq_make", "word_freq_people",
"word_freq_re", "word_freq_receive", "spam")
```

Then, we will use tidyverse to `select` only the variables from the `top_words` vector and create a new column that is the sum of the numeric columns with `mutate`:

```
# Select only columns with top words
top_df <- spam %>%
  select(all_of(top_words)) %>%
# Sum numeric columns = add column total percentage
  mutate(top_w_pct= rowSums(across(where(is.numeric))) )
```

This code snippet creates the first boxplot graphic and saves it as `g1`, using the same coding syntax that we are familiar with:

```
# Plot bar graphic for top 23 words
g1 = ggplot(top_df) +
  geom_boxplot(aes(y= spam, x= top_w_pct), fill= c("royalblue",
"coral")) +
  labs(title="How the presence of words associated with spam
e-mails impacts the classification (TOP 23)",
      subtitle= "The spam emails(1) have a higher percentage
of those words.") +
  theme_classic()
```

The same sequence of creating a column with the sum of the numeric variables and plotting a boxplot will be repeated, but this time using the entire dataset. The graphic will be saved in `g2`:

```
# Select only columns with top words
spam2 <- spam %>%
# Add total percentage
  mutate(w_pct= rowSums(across(where(is.numeric))) )


# Plot bar graphic for the entire dataset
g2 = ggplot(spam2) +
  geom_boxplot(aes(y= spam, x= w_pct), fill= c("royalblue",
"coral")) +
  labs(title="How the presence of words associated with spam
e-mails impacts the classification",
      subtitle= "The spam emails(1) have a higher percentage
of those words.") +
  theme_classic()
```

Finally, we will use the **patchwork** library to easily plot the graphics next to each other on the same figure. For that, we just need to put the g1 and g2 objects inside parentheses separated by the | sign:

```
# patchwork
(g1|g2)
```

The result is shown in *Figure 13.10*:

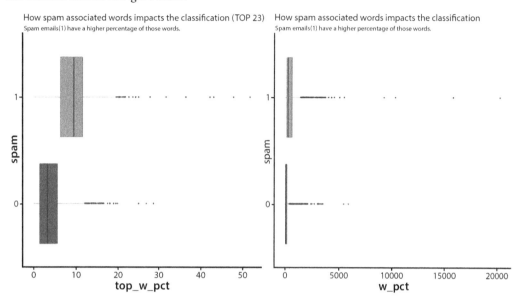

Figure 13.10 – Boxplots of spam (1) versus not spam (0) and how
they are affected by the presence of spam words

Both graphics in *Figure 13.10* show that the spam (1) emails have a much larger median of those words more likely to be associated with spam messages than the non-spam (0) messages. This is easier to detect if we consider only the top 23 words, as shown on the left side of *Figure 13.10*. But this result could be just by chance. So, let's test it statistically.

If we perform a Kolmogorov-Smirnov test, which compares both samples – spam and not spam – to test if they are from the same distribution, we will get the same conclusion that we got from the boxplots: the emails classified as spam are impacted by the number of words associated with spam:

```
# Kolmogorov-Smirnov test
yes_spam <- top_df[top_df$spam == 1,]$top_w_pct
not_spam <- top_df[top_df$spam == 0,]$top_w_pct
# Test
ks.test(yes_spam, not_spam)
```

With the hypothesis test in mind, we set the significance level to 5%. The null hypothesis will not be rejected if the p-value is more than 0.05, meaning that the samples are not statistically different. The alternative hypothesis (p-value <= 0.05) means that the samples are statistically different. The result is displayed in the following code. According to these results, the samples are different, so this difference between words will help with the classifications:

```
Warning in ks.test(yes_spam, not_spam) :
  p-value will be approximate in the presence of ties
  Two-sample Kolmogorov-Smirnov test


data:  yes_spam and not_spam
D = 0.6, p-value <0.0000000000000002
alternative hypothesis: two-sided
```

Since we know how the words impact the classification, we can start testing the characters and the uppercase impact. The code will be very similar, and it can be verified on the following GitHub page: https://tinyurl.com/mr497d8v. The result is displayed in *Figure 13.11*:

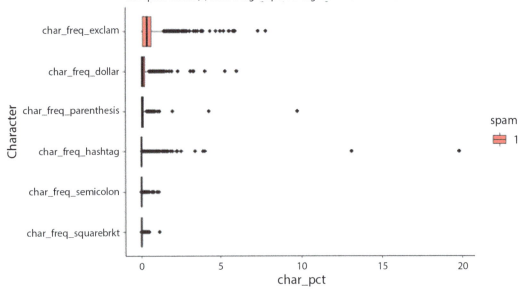

Figure 13.11 – The impact of the characters in the spam classification

Regarding the presence of uppercase letters, we can get a better idea of the impact by looking at the total number of uppercase letters in the message. This makes sense if we think about old spam messages from back in 1999, when this dataset was created, as spam emails used to be written with a lot of uppercase letters. *Figure 13.12* shows the impact of the average of uppercase letters, then the longest sequence, and finally the total number present in an email.

As mentioned previously, it is very difficult to notice this in the first two graphics, but it is clearer in the third on the right-hand side. If we perform Kolmogorov-Smirnov tests for the variables related to uppercase letters, the result is similar to the one we performed for the spam words. The p-value is very close to zero, so it doesn't give us enough statistical evidence to keep the null hypothesis. However, it does favor the alternative, thus denoting different distributions:

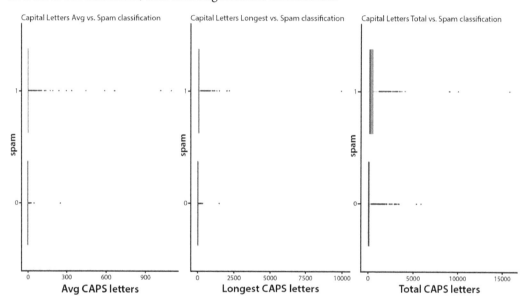

Figure 13.12 – Graphics for the uppercase letters and how it impacts the message as spam or not

By performing these explorations and looking at the visualizations, we know that the number of words, uppercase letters, and symbols are statistically different for each group in our classification. Therefore, we will continue our project by transforming the variables to make the dataset ready for modeling.

Selecting the best variables

At this point, selecting the best variables should be smooth since exploring the data gives us the answer we're looking for. When we checked the boxplots and tested the words and characters that impact the classification the most, as well as the impact of the uppercase letters, we were already making a variable selection. We should use those variables that have the highest difference between both groups so that it's easier for the algorithm to find a clearer separation between the two groups. As we have seen, 23 words maximize the difference, the number of uppercase letters, and the presence of too many symbols.

In this section, we will take the `top_words` vector, which gathers the top 23 words that have the most impact on the spam classification, as well as the exclamation, parenthesis, dollar sign, and hashtag characters and the uppercase variables and transform the dataset into a seven-variable Tibble, with six explanatory variables and one target.

The code will take the original dataset, *spam*, and use `bind_cols()` to bind the `top_w_pct` column to the sum of percentages from the words more associated with spam messages. Then, we will `select()` the target variable, spam, along with the recently created column, `top_w_pct`, plus the character variables previously mentioned and the uppercase variables. The result will be stored in the `spam_for_model` Tibble, which will be the input for the **random forest** algorithm:

```
# Creating the dataset input for modeling
# Add column top_w_pct
spam_for_model <- spam %>%
  bind_cols(top_w_pct = top_df$top_w_pct) %>%
# Select needed variables for modeling
  select(spam, top_w_pct, char_freq_exclam, char_freq_
parenthesis, char_freq_dollar, capital_run_length_total,
capital_run_length_longest)
```

The resulting dataset can be seen in *Figure 13.13*:

	spam	top_w_pct	char_freq_exclaim	char_freq_parenthesis	char_freq_dollar	capital_run_length_total	capital_run_length_longest
1	is_spam	6.74	0.778	0.000	0.000	278	61
2	is_spam	11.04	0.372	0.132	0.180	1028	101
3	is_spam	10.71	0.276	0.143	0.184	2259	485
4	is_spam	7.24	0.137	0.137	0.000	191	40
5	is_spam	7.24	0.135	0.135	0.000	191	40

Figure 13.13 – Dataset ready for the input machine learning algorithm

Now that the input dataset is ready, let's move on to the modeling portion of the project.

Modeling

Training

Now that the new dataset has been created, the next step is to replace 1 with is_spam and 0 with not_spam so that the random forest algorithm can understand that the target variable is not numeric and that it is a classification model. We can do this by using the `recode()` function within a `mutate` function:

```
# Replace the binary 1(spam) and 0(not_spam)
spam_for_model <- spam_for_model %>%
  mutate( spam= recode(spam, '1'='is_spam','0'='not_spam')    )
```

Now, it is time to separate the data into train and test subsets. The train subset is used to present the model with the patterns and the labels associated with it so that it can *study* how to classify each observation according to the patterns that occur. The test set is like a school test, where new data is presented to the trained model so that we can measure how accurate it is or how much it has learned.

As we learned during the exploration phase, this dataset is split into 60% of observations labeled as not spam and 40% as spam. In this case, we will not apply any category balancing technique as we understand that this little imbalance will not affect the result. So, let's keep going.

In the R language, there is no function for splitting the dataset into train and test sets. We want to use 80% of the data for training and 20% for testing. So, we will do that by extracting a set of random numbers from the index of the dataset.

The following code creates the random index with 80% of the row numbers. Since we know that each observation is associated with a number between 1 and 4,601, we can randomly select 3,680 (80%) to be our train set and use the rest as the test set. The randomly selected index is stored under `idx`:

```
# Create index for random train test split
number_of_rows <- nrow(spam_for_model)
idx <- sample(1:number_of_rows, size = 0.8*number_of_rows)
```

Then, we can use `idx` to split the dataset into two subsets called `train` and `test`:

```
# Split in train and test datasets
# All the observations in idx
train <- spam_for_model[idx,]
# All the observations NOT in idx
test <- spam_for_model[-idx,]
```

Next, we will quickly check if the proportions of each label are close to the numbers seen in the original dataset to make sure we don't have any biased sets. We can do that using `prop.table()`, which returns the proportions of each observation from a vector. The input for it is a `table()` function, which returns the counts of each different observation in a vector. So, combining both, we will see the percentages of each group:

```
# Label proportions of the train set
writeLines("Train set:")
prop.table( table(train$spam))
writeLines('==========================')

# Label proportions of the test set
```

```
writeLines("Test set:")
prop.table( table(test$spam) )
```

The output is as follows, where we can see that the proportions are pretty close to the original data. That's a green light to continue training:

```
 Train set:
not_spam   is_spam
   0.605     0.395

==========================
Test set:
not_spam   is_spam
   0.611     0.389
```

Finally, using the randomForest() function, we will train the machine learning model. The function receives the spam ~ . formula, meaning that the target variable, spam, will be explained by the rest, represented by the dot (.) symbol. The data for the training process is train, importance=TRUE means that we want the algorithm to calculate the most important variables, and ntree is the number of decision trees we want to create in this model:

```
# Training the model
rf <- randomForest(spam ~ ., data=train,
                    importance=TRUE, ntree= 250)
```

With that, our model has been trained. Notice that the heavy lifting is performed during the data wrangling part, where much more is coding involved. The model training process is only one line of code. Naturally, there is a lot involved in this single line, but it is important to note that the modeling part is strongly dependent on exploring and wrangling the data.

In R, the model object can be plotted using plot(rf), which inputs the model's name into the plotting function:

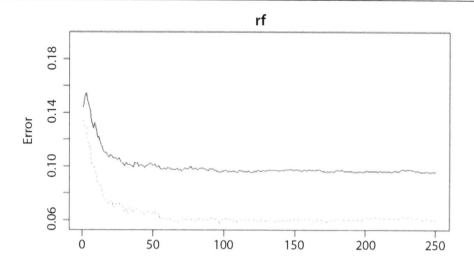

Figure 13.14 – Random forest model performance plot

Figure 13.14 shows the performance of the model. We can see that, after the 50th tree, the error stabilizes at a lower rate. Therefore, in this case, it does not need much more than that to recognize the patterns from the data.

Another nice graphic to look at is the variable importance plot:

```
# Variable importance plot
varImpPlot(rf)
```

The preceding code will result in the following output:

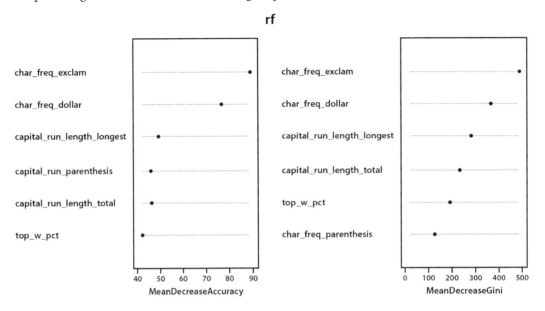

Figure 13.15 – ROC curve

Figure 13.15 shows the importance of each variable regarding the model's classifications. Observe that two metrics can be used: the mean decrease accuracy and the Gini index.

In just a few words, these two measurements are how the **random forest** algorithm can decide if a certain decision brings a better result to the model or not. For this library, the default for classification models is the Gini index. They are displayed by order of importance, so the most important variables for the classifications in our model are those related to the characters.

Let's test the model and see how accurate it is.

Testing and evaluating the model

There are many good metrics for testing a classification model. The most used metrics are **accuracy**, **confusion matrix**, and **receiver operating characteristic** (**ROC**) curve. We will go over a couple of them in this project.

First, let's check the **accuracy** of the model. In simple terms, out of the total number of classifications, we can see how many of those the model predicted correctly. That measure is created by comparing the model's predictions and the true classifications from the test dataset. We'll use the `predict ()` function for this while using the `rf` model object and the `test` set as input to make the predictions:

```
# Predictions
preds <- predict(rf, test)
```

Using the `preds` object, we can create a confusion matrix, using a function with the same name from the **caret** library. The inputs are the predictions and the labels (`test$spam`) from the test set:

```
# Confusion Matrix
confusionMatrix(preds, test$spam)
```

The result contains a lot of good information. In the beginning, it shows the **confusion matrix**, which compares the prediction and the real data. So, it takes the target variable labels of not_spam and is_spam as reference columns. Then, it takes the predicted labels of not_spam and is_spam and places those as lines. We can see a 2x2 matrix where the main diagonal (positions 1,1 and 2,2) shows the correct predictions. (2,1) is the **false positive** or the **Type 1** error. The position (1,2) is the **false negative** or **Type 2** error.

In the sequence, we can see the accuracy of the model. It shows that out of the 100 predictions, there will be around 7 errors (93% accuracy), with a confidence interval floating between 6 to 9 errors. We can also see some other statistical tests and measurements:

```
Confusion Matrix and Statistics
          Reference
Prediction not_spam is_spam
  not_spam      540       40
  is_spam        23      318

                Accuracy : 0.932
                  95% CI : (0.913, 0.947)
     No Information Rate : 0.611
     P-Value [Acc > NIR] : <0.0000000000000002

                   Kappa : 0.855
  Mcnemar's Test P-Value : 0.0438
             Sensitivity : 0.959
             Specificity : 0.888
          Pos Pred Value : 0.931
```

```
           Neg Pred Value : 0.933
              Prevalence : 0.611
          Detection Rate : 0.586
    Detection Prevalence : 0.630
       Balanced Accuracy : 0.924


          'Positive' Class : not_spam
```

The numbers look good. But we can still look at the **ROC** curve to check the performance based on the rate of true positives and false positives. Let's start by creating the prediction probabilities with the predict() function:

```
# Predictions probability of not_spam
predictions <- data.frame(predict(rf, test, type='prob'))
```

Then, using the ROCR library, we will create a prediction object, providing the prediction function with the positive class, predictions$not_spam, and the real labels of *spam* or *not spam* by observation from the test set:

```
# Predictions for ROC
pred_roc <- prediction(predictions$not_spam, test$spam)
```

Next, using the performance() function, we will input the prediction object, pred_roc, the labels we want for the axes (tpr is the true positive rate and fpr is the false positive rate), and abline to draw a diagonal that separates the graphic into two, representing the 50/50 rate:

```
# ROC curve
roc <- performance(pred_roc,"tpr","fpr")
plot(roc, colorize = T, lwd = 2)
abline(0.0, 1.0)
```

The result is shown in *Figure 13.16*:

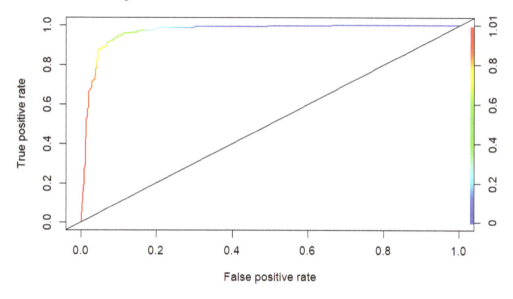

Figure 13.16 – ROC curve

Interpreting *Figure 13.16* is easier than it looks: the more the graphic goes to the upper left-hand side, the better the model, given that it is approximating from 100% of the true positives, where the classification was correctly performed. It also shows that the rate of true positives is growing much faster than the false positives, so we will see more correct predictions. On the flip side, seeing a line closer to the middle diagonal is like a model guessing the result, since there is a 50% of chance of predicting each class. Below that line, the model did worse than guessing and missed any patterns in the data.

Next, let's put the model to work by entering some new text into it.

Predicting

After evaluating the model, it is time to test the model with some random text created with and without spam words to see how the model will behave. To input a value for prediction, the new observations must be in the same format as the data used for training and testing. It must have the same explanatory and response variables as the DataFrame from *Figure 13.13*. This means that every transformation made during the data preparation phase will have to be repeated before we can input text for prediction. For this, we will use a custom-created function called `prepare_input()`, which can be seen on the GitHub page for this chapter. This function takes a string of text and the `spam_words` vector as input. Then, it reads the text to count the quantities of *!*, *$*, *()*, uppercase letters, the longest sequence of uppercase, and words in the spam list, returning a DataFrame for input in the RF model. The returned object is in the same format as the DataFrame from *Figure 13.13*.

The sample texts to be used are as follows. The first is spam text, full of uppercase text, spam words, and characters:

```
text1 <- "SALE!! SALE!! SALE!! SUPER SALEEEE!! This is one of
the best sales of the year! More than #3000# products with
discounts up to $500 off!! Visit our page and Save $$$ now!
Order your product NOW (here) and get one for free!"
```

The other is a regular message, which contains some uppercase text and a couple of spam words such as *you*, *your*, *will*, and *received*, to see how the model will behave in this case:

```
text2 <- "DEAR MR. JOHN, You will find enclosed the file we
talked about during your meeting earlier today. The attachment
received here is also available in our web site at this
address: www.DUMMYSITE.com."
```

Next, we will create a vector with the top spam words:

```
spam_words <- c("you", "your", "will", "free", "our", "all",
"mail", "email", "business", "remove", "000", "font", "money",
"internet", "credit", "over", "order", "3d", "address", "make",
"people", "re", "receive")
```

Before predicting the result of the first piece of text, we input text1 into the custom prepare_input () function and store it in the input variable. Then, we use predict, which receives the model's rf, the input variable, and the type= "prob" argument, to return the probabilities. The code is wrapped by a data.frame function to return a formatted result. See the code in the sequence:

```
# Predicting text 1
input <- prepare_input(text1, spam_words)
#Prediction
data.frame(predict(rf, input, type="prob"))
```

The result of the prediction is shown in the sequence:

Description: df [1 x 2]

	not_spam <dbl>	is_spam <dbl>
1	0.34	0.66

1 row

Figure 13.17 – Prediction of the spam text

Figure 13.17 shows a correct prediction, with a 66% probability that the text is a spam message. Let's try the same for `text2`. First, format the input:

```
# Predicting text 2
input <- prepare_input(text2, spam_words = spam_words)
```

Then, predict the probabilities:

```
#Prediction
data.frame(predict(rf, input, type="prob"))
```

Description: **df [1 x 2]**

	not_spam <dbl>	is_spam <dbl>
1	0.704	0.296

1 row

Figure 13.18 – Prediction of the not-spam text

Figure 13.18 also shows a correct prediction, with a 70% probability that the text is not a spam message, even with the spam words and uppercase letters that we added to try to fool the model.

Finally, we must save this model for later deployment. To save a model, we can use the `saveRDS()` function and input the model's name and the name of the output filename:

```
# Saving the model
saveRDS(rf, "rf_model.rds")
```

In the next chapter, we will deploy the model in a **Shiny** application that we are going to build with R coding. A deployment is when we get the model to do its job – that is, to make predictions in the real world – while receiving input from a user via an internet application and returning the prediction.

Summary

In this chapter, we created an end-to-end machine learning project. We started by studying some basic machine learning concepts to put us in sync. Then, we understood what was needed for the main goal of the project. First, we must understand the problem and know where we want to go so that the solution becomes clearer. In this case, our client was a digital marketing company that wanted to reduce the risk of their messages ending up in their spam filter, so we had to create a classification model to predict the probability of a message being marked as spam or not spam.

We loaded a dataset from UCI, which brought up some words and characters associated with spam messages and their percentage in the email. Then, we studied the data and created some visualizations to learn which elements were more likely to be classified as spam. Out of those, we created a new dataset with just six explanatory variables, reducing it from the original 57 columns.

Next, we trained and tested the model, which resulted in a machine learning model that we created with the **random forest** algorithm and performed with an average of 93% accuracy. So, out of 100 messages, only seven are at risk of being incorrectly classified.

In the next chapter, we'll build the application to deploy the model.

Exercises

1. What are the two types of models in machine learning?

2. What are the three learning methods of machine learning algorithms?

3. What is the importance of data wrangling for modeling data?

4. Before creating a model, what is the split we must do with the dataset?

5. What are some of the metrics we can use to evaluate a classification model?

6. Explore the UCI database and choose a dataset to create another model.

Further reading

*   Spambase dataset, created by Mark Hopkins, Erik Reeber, George Forman, and Jaap Suermondt and donated by George Forman: `https://archive.ics.uci.edu/ml/datasets/spambase`

*   Supervised, unsupervised, and reinforcement learning algorithms: `https://tinyurl.com/sxcr9xpj`

*   Bagging ML models: `https://en.wikipedia.org/wiki/Bootstrap_aggregating`

*   Random Forest: `https://www.r-bloggers.com/2021/04/random-forest-in-r/`

*   Gini/mean decrease accuracy: `https://tinyurl.com/272auchs`

*   Gini index for the random forest algorithm: `https://tinyurl.com/42p8umhb`

*   ROC curve: `https://en.wikipedia.org/wiki/Receiver_operating_characteristic`

*   Difference Between Classification and Regression in Machine Learning: `https://tinyurl.com/mr2j436a`

*   Creating side-by-side plots with patchwork: `https://tinyurl.com/mecmhf8k`

*   R code GitHub page for this chapter: `https://tinyurl.com/mr497d8v`

14

Build an Application with Shiny in R

We have finally reached the last chapter. We have learned so much during this book, and we have been able to consolidate our new knowledge with the machine learning project we created in the last chapter. Now, it is time to put that model in production, making it available for the final user.

Putting a model into production is nothing more than taking it from the environment where it was created and trained, usually an internal environment in a company or even a data scientist's local machine, and making it available in an application, serving the purpose it was created for.

In this project, putting the model into production entails creating a web app with the **Shiny** library, embedding the model into it, and making it available to receive textual input from users. It will predict the probability of that text being spam or not.

This is our plan for this final chapter:

- Learning the basics of Shiny
- Creating an application
- Deploying the application on the web

Technical requirements

All the code can be found in the book's GitHub repository: `https://github.com/PacktPublishing/Data-Wrangling-with-R/tree/main/Part4/Chapter14`.

The libraries we need are as follows:

```
library(shiny)
library(shinythemes)
library(plotly)
library(tidyverse)
library(randomForest)
```

Learning the basics of Shiny

Shiny for R was launched in 2012 by Winston Chang as a package to help R developers to create interactive UIs for R scripts. This way, the developers can not only create models, but also share them with other people via an application. Currently, Shiny is on version 1.7.0 and is more robust than ever, expanding its horizons to other programming languages.

As a side note, let's define an application, since we will use this term a lot in this chapter. An application, or just app, is software created via programming language to solve a problem. It performs a task, such as text editing, web browsing, or translation. A web application is just an app on the internet, such as the many websites where we can do things and interact, such as document creation, photo editing, or design creation.

Among the options available on the market, Shiny stands out for being a complete solution for the developer, offering many gadgets that can be added to applications, as well as integrating well with other famous players such as **plotly** for interactive graphics.

Get started

While this is not a book about applications, it is useful to have at least a high-level view of how things work under the hood of a Shiny application. The installation goes just like any other R package, using `install.packages("shiny")`.

The library has a built-in example that we will go over in this introduction, so I suggest that we run the next code snippet to look at the example app while we learn about its components:

```
library(shiny)
runExample("01_hello")
```

Once we run the application, we see the screen shown in *Figure 14.1*.

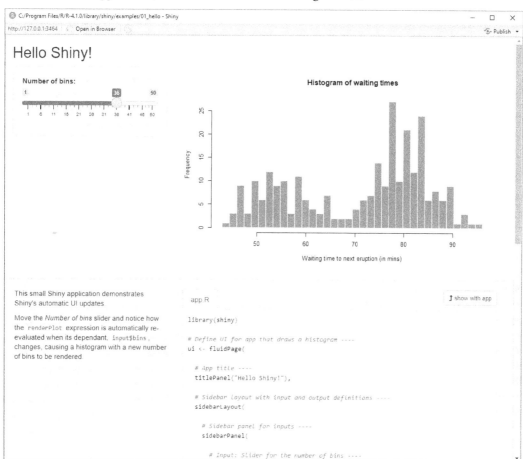

Figure 14.1 – Shiny application built-in example

Figure 14.1 is displaying the app `01_hello`, a simple example of how Shiny applications work. Basically, Shiny apps have three main components: the **UI** and the **server function**, plus a **function** to call the app to load. Let's learn more about these components:

- UI: In the app `01_hello`, whatever you see in the top part of the screen (above the gray area) is the UI. As the name suggests, this is where the interaction between a user and the application happens. Thus, this is where the developer will add visual elements, such as text, tables, images, and graphics, as well as interactive gadgets such as sliders and buttons.

- Server function: For the UI to work, something must feed it. That is the role of the server function. This component is the backstage of the app, which the client will not see. This is where the calculations will happen, along with the creation of tables and graphics, which are then rendered by the UI.

- Call function: The last piece is a call function, which works like a connector. It calls the interface and the server functions and makes them work together, loading the app to the end user.

Here is the backbone code of a Shiny app:

```
# User Interface
ui <- fluidPage(   )
#Server functions
server <- function(input, output) {   }
# Call function
shinyApp(ui = ui, server = server)
```

In summary, these components can be described as the frontend (UI) and the backend (server function), with a calling function that loads the application.

Going back to *Figure 14.1* once more, the instructional text in the app states that if you move the slider on the top left-hand side of the screen, the histogram graphic will change accordingly, increasing or decreasing the number of bins. The slider, the text, and the rendering of the graphic are coded in the UI. The histogram plot and the ability to change the number of bins based on the slider's input, which is controlled by the server function.

Basic functions

I strongly recommend that you go over the tutorials on the Shiny web page. The content is very clear, complete, and friendly to most levels of programming knowledge. In the following table, we will study some of the main functions used by the library, because I believe it can give you a head start when reading any Shiny application code and it will help you to consolidate your learning as well.

Funtion	Description
`fluid()`	Creates a page where the content will be displated to the user
`navbarPage()`	Creates a menu with tabs for navigation
`tabPanel()`	Creates a menu with tabs for navigation
`mainPanel()`	Add content to the main panel of a page or tab
`sidebarPanel()`	Add content to a sidebar panel of a page or tab
`p(), h1(), h2(), a(), img(), div()`	Functions for HTML-equivalent notation for textual content, such as title sizes, reference links, images, and divisions
`column()`	Sets up content in columns
`selectInout(), radioButtons(), numericInput(), sliderInput(), textInput, actionButton()`	Functions to add interactive gadgets to the UI
`plotOutput(), tableOutput(), tableOutput()`	Functions to add graphics, tables, and text the are created by the server functions in the backend
`renderPlot(), renderTable(), renderText()`	Functions to be used in the server functions portion that pair with the output functions

Figure 14.2 – Basic Shiny functions

Let's move on to build our application with Shiny to deploy the model created in *Chapter 13*.

Creating an application

We created a classification model that is able to estimate the probability of any text being classified as spam or not spam, based on the most common spam words and characters from the *Spambase* dataset. However, if we never add that model to a tool where a person can input text, the likelihood is that the model will become useless. So, the solution is to deploy it, embedding the classifier in a web application. Let's define our project next.

The project

The project for this last chapter is described in the following bullet points:

- **Problem**: Create an interactive application able to deploy a machine learning model to the web.
- **Description**: The tool will be able to receive textual input, transform the data to a data frame that will feed the machine learning random forest classifier. The model predicts the probability that a text message is spam or not.
- **Tools**: Shiny library and RStudio.

Coding

Now that our project is clear, let's get our hands dirty and start coding. The first step is loading the libraries needed for this work. We are going to need the **tidyverse** for data wrangling, Shiny and Shiny **Themes** for the app creation, **plotly** for interactive graphics, and **randomForest** to run our model:

```
library(shiny)
library(shinythemes)
library(plotly)
library(tidyverse)
library(randomForest)
```

The model is already created, trained, and we saved it. So, we must load it to the app to predict the results. Following is a single line of code using the readRDS() function:

```
model <- readRDS("rf_model.rds")
```

The code will create a vector with the top influencing spam words that we determined during the exploratory analysis in *Chapter 13*:

```
# Define spam words
spam_words <- c("you", "your", "will", "free", "our", "all",
"mail", "email", "business", "remove", "000", "font", "money",
"internet", "credit", "over", "order", "3d", "address", "make",
"make", "people", "re", "receive")
```

Additionally, there is a custom function (prepare_input) that I am not adding here for the sake of space. It is an extensive custom function that can be seen on the book's GitHub page (https://tinyurl.com/2p838wnr). It is all properly commented and the functions used to create it are all covered in this book.

UI

Next, we will start creating the UI, which is the part of the application the user can interact with. Notice that the entire code for this app is long, with almost 250 lines, so we will break it into chunks that are small enough to be understandable, and part of it will be omitted for not adding essential information.

We begin by adding a page with `fluidPage()` and defining a theme with `shinytheme()`. Then we add `navbarPage()`, which is the top bar that allows us to navigate through the two tabs of the project:

```
# Creating the UI
ui <- fluidPage(theme = shinytheme("united"),
                # Panel with all the tabs on top of the pages
                navbarPage(
                    theme = "united",
                    title="Data Wrangling with R",
```

At this point, we are starting the code for the **About** tab, where information about the project, the dataset used, the author, and the model are displayed.

> **Hint**
> Since this code is long and there are many parenthesis and curly brackets, I encourage you to always use a comment after the end of a function to know what part of the code you are closing. For example, observe that I add comments like `#fluidRow-About`, `#mainPanel-About` and `#tabPanel-About`.

When we come back to this code, a few weeks later, it becomes easy to know what function is ending at that point.

```
                # Tab About
                #----------------------------------
                tabPanel("About the Project",
                        mainPanel(fluidRow(
                            <<<content here>>> ) # fluidRow-
    About
            ) # mainPanel-About
        ), #tabPanel-About
```

The second tab is called **SPAM Classifier**. This is where the interaction will happen. Observe that there are many other functions within it, such as `textAreaInput()`, where the user will write the text to be classified; `submitButton()`, used to send the input for classification; and `tableOutput()`

and `plotlyOutput()`, which will work with the rendering functions from the server side to put the results in a table and in a graphic figure:

```
# Tab SPAM CLASSIFIER
        tabPanel("SPAM Classifier",
                mainPanel(fluidRow(
                  #text Input
                  textAreaInput(inputId = "text",
  NULL, " ", width = "1000px", height="120px"),
                  submitButton(text= "Submit"),
                    h4( tableOutput("prediction") )
                      ),# column1
                  plotlyOutput(outputId =
  measurements")

                    )#column2
          ) # fluidRow-spam_classifier
          ) # mainPanel-spam_classifier
     ) #tabPanel-spam_classifier
```

With that, our UI is ready. But the app will not work with just that portion of the code. As we learned before, this is the frontend: what we see of the app. But a frontend is always fed by a function on the backend. That is what we will code next.

Server function

On the server side, we will create two functions that will feed the outputs on the UI side. But first, we must create the backbone code for the server side:

```
# Creating a server function
server <- function(input, output) { }
```

The first function that will go inside the server is `output$prediction`. It takes the text from `textAreaInput()` on the UI side as input and runs the custom `prepare_input()` function created for this project. Then, the resultant variable, `datatext`, feeds the `predict()` function, together with the model, and returns the probabilities. These numbers are going to be displayed by `tableOutput()` on the UI side:

```
# Piece of code for the Prediction
  output$prediction <- renderTable({
    # Prepare data for input in the model
```

```
datatext <- prepare_input(input$text)

# Predict
prediction <- predict(model, datatext, type="prob")
data.frame(prediction*100)
}#end if
}) #output prediction
```

The second function, called `output$measurements`, creates a bar plot that will go into the `plotlyOutput()` on the UI side. Similarly, it also takes the textual input and runs the `prepare_input()` custom function, storing the result in `datatext`, which is a data frame. Next, `datatext` is transposed and converted to a `data.frame` object. Then we create a vector named `measurements` to be the names of the categories for each bar:

```
# Piece of code for the measurements graphic
output$measurements <- renderPlotly({
  # Prepare data as a dataset
  datatext <- prepare_input(input$text) #receive text input
  datatext <- data.frame(t(datatext)) #transpose datatext
  colnames(datatext) <- "values" #change column name
  datatext <- as.data.frame(datatext) #convert to data.frame
object
  # Names for the categories
  measurements <- c("Spam words", "Presence of !!!","Presence
of ( )", "Presence of $$$", "Total UPPER", "Longest UPPER")
```

Finally, we plot the bar plot with the `plot_ly()` function, inserting the dataset (`data`), the x and y axes, and the graphic's `type`, `color`, and transparency (`alpha`) as arguments. In addition to that, there is another `layout()` function for setting up the width, height, and *x*-axis range for the plot:

```
# Create graphic
g <- plot_ly(data = datatext, x = ~values, y =
measurements, type = "bar", name="Bar", alpha = 0.85, color=
"darkorange")
g <- g %>% layout(width = 500, height = 200,
                      xaxis=list(range=c(0,100)) )
plotly_build(g)
}#end if
}) #output bar graphic
```

That is all for the server function. Our backend is complete now, with two functions that will feed the `tableOutput()` and the `plotlyOutput()` functions on the UI side.

Call function

Once all that coding is done, the last piece missing is the call function. That function is a single line of code that tells the R language where to look for the UI and server sides. Given that we named our UI as `ui` and the server function as `server`, the code snippet is as follows:

```
shinyApp(ui = ui, server = server)
```

The application

We are good now. I bet you are very anxious to see the result. To be able to run the app on your local system, you must save the file with the name `app.R`. With that done, the run button will appear now as **Run App**.

When the **Run App** button is clicked, RStudio will create a local server on your machine and run the app. *Figure 14.3* and *Figure 14.4* display the resulting application:

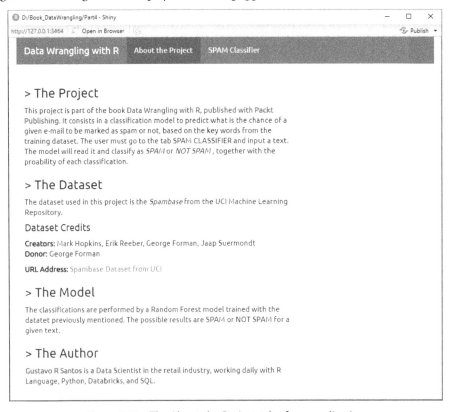

Figure 14.3 – The About the Project tab of our application

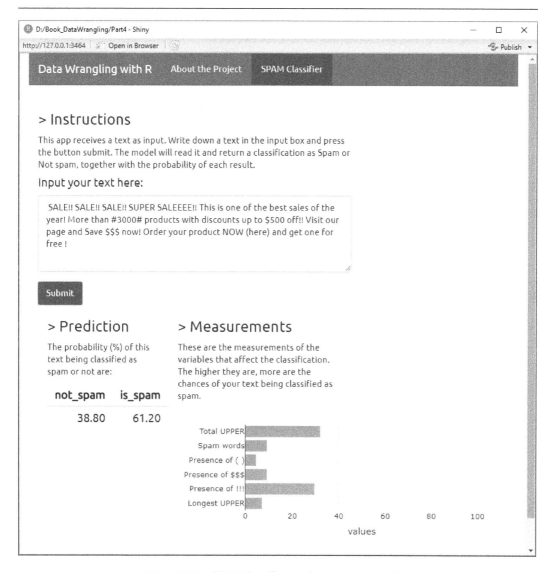

Figure 14.4 – SPAM Classifier tab from our application

The model is working properly locally. However, we still have one step for the final deployment: to put it online, on the internet.

Deploying the application on the web

Working with Shiny and RStudio makes it really easy to deploy an app. Once it is ready for deployment, all we need to do is click the blue **Publish Document…** button in the top right-hand corner of the source code window in RStudio, right next to the **Run App** button, as shown in *Figure 14.5*.

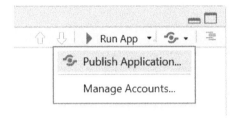

Figure 14.5 – Publish Application option for Shiny apps in RStudio

A pop-up window will appear, asking you to connect RStudio to an account. Hit the **Next** button. See *Figure 14.6*.

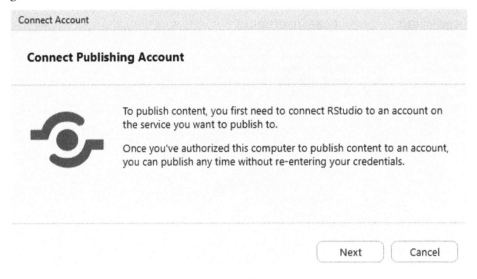

Figure 14.6 – Publishing options

The next step is to select where you want to publish it. *ShinyApps.io* has a free version for students. You can choose it, as shown in *Figure 14.7*.

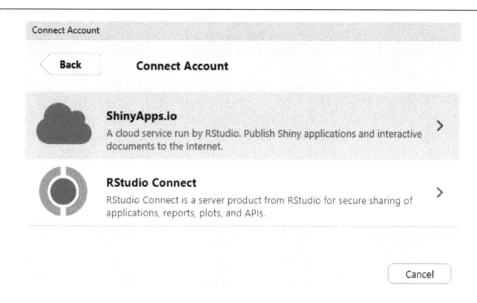

Figure 14.7 – Options for publishing an application

To connect to the account, you must register on the website and get a token to connect to *ShinyApps*. Follow the instructions and paste the result in the input area. After that, click **Connect Account**, as displayed in *Figure 14.8*.

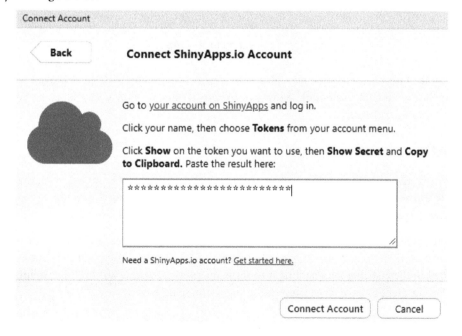

Figure 14.8 – Connecting to ShinyApps.io

Once your account is connected, the next window that appears is the following one, shown in *Figure 14.9*. Here, we must check the boxes to select the files used to build the application. Select **app.R** and the model **rf_model.rds**, then click **Publish**.

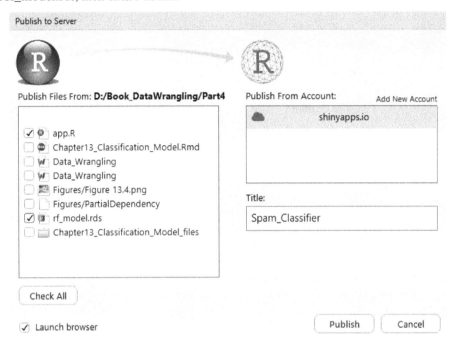

Figure 14.9 – Connecting to ShinyApps.io

It takes a couple of minutes, but soon enough a web browser will pop up with the deployed app. It is done. The application is deployed online, ready to be used by anyone with the URL address `https://tinyurl.com/5n6vbvrn`.

Summary

We have reached the end of this book, and we closed it by learning about a fantastic tool: the Shiny library for R. Shiny is a tool for building interactive applications, enabling R developers to deploy and show their work online.

In this chapter, we learned the basics of the Shiny library, how to get started, and the most common functions, and we then turned our efforts to build an application that wrapped the random forest model trained in the last chapter, enabling it to serve the purpose that it was built for: helping our digital marketing client to have fewer messages going to the spam box.

The project was based on patterns from an open dataset from the UCI repository; therefore, it will not be applicable to any specific emails or any messages, but more specifically to the patterns learned from that dataset, to solve the problem from our hypothetical client, which was our goal.

We ended the chapter by deploying the app online using the `ShinyApps.io` free version.

Exercises

1. What is an application?

2. What is Shiny?

3. What are the three main components of Shiny?

4. What does it mean to deploy an app?

Further reading

*   Shiny Get Started page: `https://shiny.rstudio.com/tutorial/#get-started`

*   Shiny course: `https://rstudio-education.github.io/shiny-course/`

*   Application: `https://en.wikipedia.org/wiki/Application_software`

*   ShinyApps.io website: `https://www.shinyapps.io/`

*   Application created in this chapter: `https://tinyurl.com/5n6vbvrn`

*   R code GitHub page for this chapter: `https://tinyurl.com/2p838wnr`

Conclusion

Every journey reaches an end. And here, we end ours, after fourteen chapters of content. First, I would like to thank you for buying and reading this book. It took me a good amount of research and work to put together a book that I expect will help many current and future data professionals.

Wrangling data is an essential part of any project, as it will determine the quality of the output. Trust me, there will be no project where you will face *ready-to-go* datasets, so knowing how to clean and transform your data is a must-have skill in data science.

In my job, I am constantly challenged to present data in many different ways, so I can say that I use the knowledge from this book on a daily basis, and I hope you can do that too from now on. If you already knew part of what we studied, I hope you could capture some hints that will enhance your skills.

Remember, once you know what needs to be the outcome, you can focus on the best transformations. Always keep the desired output in mind. Then, think about the steps that will get you to that result, and then you can decide what the best transformation is for each task.

I encourage you to review the code that accompanies this book in GitHub and practice a lot to internalize your new learnings. Other than that, I also wish you success in your future endeavors as a data professional.

Cheers!

References

Altair, 2022. What is Data Wrangling? Available at < `https://www.altair.com/what-is-data-wrangling/#:~:text=Data%20wrangling%20is%20the%20process,to%20be%20organized%20for%20analysis.`>. Access on: 20 June 2022.

Bruce, Peter; Bruce, Andrew,2019. Practical Statistics for Data Scientists: 50 Essential Concepts. 1ed. O'Reilly, Sebastopol, CA, EUA.

The Comprehensive R Archive Network [CRAN], 2022. Introduction to data.table. Available at < `https://cran.r-project.org/web/packages/data.table/vignettes/datatable-intro.html`>. Access on: 30 June 2022.

Data Novia, 2018. GGPLOT TITLE, SUBTITLE AND CAPTION. Available at < `https://www.datanovia.com/en/blog/ggplot-title-subtitle-and-caption/`>. Access on: 26 August 2022.

Data Science Glossary, 2017. Data Science Glossary. Available at < `https://www.datascienceglossary.org/`>. Access on: 20 June 2022.

Express Analytics, 2022. What Is Data Wrangling? What are the Steps in Data Wrangling? Available at < `https://www.expressanalytics.com/blog/what-is-data-wrangling-what-are-the-steps-in-data-wrangling/`>. Access on: 20 June 2022.

Géron, A. 2019. Hands-on Machine Learning with Scikit-Learn, Keras and Tensorflow. 2ed. O'Reilly, Sebastopol, CA, EUA.

Klein, Grady; Dabney, Alan, 2013. The Cartoon Introduction to Statistics. 1ed. Hill and Wang, New York, NY, EUA.0

OSDC, 2021. Data Wrangling: How to Prepare Your Data. Available at 00< `https://opendatascience.com/data-wrangling-how-to-prepare-your-data/`>. Access on: 25 June 2022.

Quantum, 2022. Data Science project management methodologies. Available at < `https://medium.datadriveninvestor.com/data-science-project-management-methodologies-f6913c6b29eb`>. Access on: 30 June 2022.

R-Bloggers.com, 2021. Handling missing values in R. Available at < `https://www.r-bloggers.com/2021/04/handling-missing-values-in-r/`>. Access on: 20 July 2022.

Shiny.com, 2022. Shiny official documentation. Available at < `https://shiny.rstudio.com/tutorial/`>. Access on: 07 September 2022.

Stack Overflow, 2022. bind_rows() creates duplicate of each dataframe when binding them in R. Available at < `https://stackoverflow.com/questions/72365708/bind-rows-creates-duplicate-of-each-dataframe-when-binding-them-in-r`>. Access on: 26 July 2022.

Stack Overflow, 2013. Fetching date/time from string in R [closed]. Available at <https://stackoverflow.com/questions/16986968/fetching-date-time-from-string-in-r>. Access on: 15 July 2022.

Stack Overflow, 2014. ggplot2 line chart gives "geom_path: Each group consists of only one observation. Do you need to adjust the group aesthetic?" Available at <https://stackoverflow.com/questions/27082601/ggplot2-line-chart-gives-geom-path-each-group-consist-of-only-one-observation>. Access on: 27 August 2022.

Stack Overflow, 2017. How to plot 3D scatter diagram using ggplot? Available at <https://stackoverflow.com/questions/45052188/how-to-plot-3d-scatter-diagram-using-ggplot>. Access on: 30 August 2022.

Tidyverse.org, 2022. Dplyr official documentation. Available at <https://dplyr.tidyverse.org/index.html>. Access on: 10 September 2022.

Tidyverse.org, 2022. Ggplot2 official documentation. Available at <https://ggplot2.tidyverse.org/>. Access on: 25 August 2022.

Tidyverse.org, 2022. Readr official documentation. Available at <https://readr.tidyverse.org/>. Access on: 30 June 2022.

Tidyverse.org, 2022. Stringr official documentation. Available at <https://stringr.tidyverse.org/>. Access on: 04 July 2022.

Tidyverse.org, 2022. Tidyr official documentation. Available at <https://tidyr.tidyverse.org/>. Access on: 25 July 2022.

Wei, Taiyun; Simko, Viliam, 2021. An Introduction to corrplot Package. Available at <https://cran.r-project.org/web/packages/corrplot/vignettes/corrplot-intro.html>. Access on: 23 August 2022.

Wickham, Hadley, 2014. Tidy Data. The Journal of Statistical Software, vol. 59.

Wickham, Hadley; Grolemund, Garret, 2017. R for Data Science. 1ed. O'Reilly, Sebastopol, CA, EUA.

Index

Symbols

3D plots 282-287

A

analysis report 238, 239
anti-join 205, 206
API
  data, obtaining from 47-49
application
  deploying, on web 344-346
application, building with Shiny 337
  call function, creating 342
  coding 338
  project 338
  running 342, 343
  server function, creating 340-342
  user interface (UI), creating 339, 340
apply functions 119
  apply 120
  lapply 120, 121
  sapply 121, 122
  tapply 122
arithmetic operations
  with datetime 136
artificial intelligence (AI) algorithms 309

B

bar plot 64, 254-259
  creating 64-67
Base-R 52
binding 172, 173
bootstrap aggregating 309
boxplot 35, 56, 251, 252
  creating 56-59
business intelligence (BI) tools 291

C

caret library 310, 327
Census Income dataset
  reference link 156
central limit theorem (CLT) 235
Central Standard Time (CST) 139
classification model 306
  evaluating 327-329
  testing, metrics 326
  text, predicting 329-331
coefficient of variation (CV) 219
comma separated values (CSV) file
  data, customizing while import 21-23
  functions, comparing to load 20
  loading, to R 17-19

Packt.com

Subscribe to our online digital library for full access to over 7,000 books and videos, as well as industry leading tools to help you plan your personal development and advance your career. For more information, please visit our website.

Why subscribe?

- Spend less time learning and more time coding with practical eBooks and Videos from over 4,000 industry professionals

- Improve your learning with Skill Plans built especially for you

- Get a free eBook or video every month

- Fully searchable for easy access to vital information

- Copy and paste, print, and bookmark content

Did you know that Packt offers eBook versions of every book published, with PDF and ePub files available? You can upgrade to the eBook version at packt.com and as a print book customer, you are entitled to a discount on the eBook copy. Get in touch with us at customercare@packtpub.com for more details.

At www.packt.com, you can also read a collection of free technical articles, sign up for a range of free newsletters, and receive exclusive discounts and offers on Packt books and eBooks.

Other Books You May Enjoy

If you enjoyed this book, you may be interested in these other books by Packt:

SQL for Data Analytics - Third Edition

Jun Shan, Matt Goldwasser, Upom Malik, Benjamin Johnston

ISBN: 978-1-80181-287-0

- Use SQL to clean, prepare, and combine different datasets
- Aggregate basic statistics using GROUP BY clauses
- Perform advanced statistical calculations using a WINDOW function
- Import data into a database to combine with other tables
- Export SQL query results into various sources
- Analyze special data types in SQL, including geospatial, date/time, and JSON data
- Optimize queries and automate tasks
- Think about data problems and find answers using SQL

Hands-On Data Preprocessing in Python

Roy Jafari

ISBN: 978-1-80107-213-7

- Use Python to perform analytics functions on your data
- Understand the role of databases and how to effectively pull data from databases
- Perform data preprocessing steps defined by your analytics goals
- Recognize and resolve data integration challenges
- Identify the need for data reduction and execute it
- Detect opportunities to improve analytics with data transformation

Packt is searching for authors like you

If you're interested in becoming an author for Packt, please visit `authors.packtpub.com` and apply today. We have worked with thousands of developers and tech professionals, just like you, to help them share their insight with the global tech community. You can make a general application, apply for a specific hot topic that we are recruiting an author for, or submit your own idea.

Share Your Thoughts

Now you've finished *Data Wrangling with R*, we'd love to hear your thoughts! Scan the QR code below to go straight to the Amazon review page for this book and share your feedback or leave a review on the site that you purchased it from.

`https://packt.link/r/1804611476`

Your review is important to us and the tech community and will help us make sure we're delivering excellent quality content.

Download a free PDF copy of this book

Thanks for purchasing this book!

Do you like to read on the go but are unable to carry your print books everywhere?

Is your eBook purchase not compatible with the device of your choice?

Don't worry, now with every Packt book you get a DRM-free PDF version of that book at no cost.

Read anywhere, any place, on any device. Search, copy, and paste code from your favorite technical books directly into your application.

The perks don't stop there, you can get exclusive access to discounts, newsletters, and great free content in your inbox daily

Follow these simple steps to get the benefits:

1. Scan the QR code or visit the link below

https://packt.link/free-ebook/9781803235400

2. Submit your proof of purchase
3. That's it! We'll send your free PDF and other benefits to your email directly

www.ingramcontent.com/pod-product-compliance
Lightning Source LLC
Chambersburg PA
CBHW062046050326
40690CB00016B/2994